GEOMETRY AND ALGEBRA
IN
ANCIENT CIVILIZATIONS

B. L. van der Waerden

Geometry and Algebra
in
Ancient Civilizations

With 98 Figures

Springer-Verlag
Berlin Heidelberg New York Tokyo
1983

Prof. Dr. B. L. van der Waerden
Mathematisches Institut der Universität Zürich

QA
151
W34

Cover illustration

One of Tai Chen's illustrations of the "Nine Chapters of the Mathematical Art", explaining Liu Hui's method of measuring the circle (see pages 196–199). Reproduced from Joseph Needham: Science and Civilization in China, Volume 3, p. 29.

ISBN 3-540-12159-5 Springer-Verlag Berlin Heidelberg New York Tokyo
ISBN 0-387-12159-5 Springer-Verlag New York Heidelberg Berlin Tokyo

Library of Congress Cataloging in Publication Data
Waerden, B. L. (Bartel Leenert) van der, 1903–
Geometry and algebra in ancient civilizations.
Includes index.
Contents:
1. Algebra – History. I. Title.
QA151.W34. 1983. 512'.009. 83-501
ISBN 0-387-12159-5 (U.S.)

© Springer-Verlag Berlin Heidelberg 1983
Printed in Germany

Typesetting and printing: Zechnersche Buchdruckerei, Speyer
Bookbinding: J. Schäffer OHG, Grünstadt
2141/3140-543210

Preface

Originally, my intention was to write a "History of Algebra", in two or three volumes. In preparing the first volume I saw that in ancient civilizations geometry and algebra cannot well be separated: more and more sections on ancient geometry were added. Hence the new title of the book: "Geometry and Algebra in Ancient Civilizations". A subsequent volume on the history of modern algebra is in preparation. It will deal mainly with field theory, Galois theory and theory of groups.

I want to express my deeply felt gratitude to all those who helped me in shaping this volume. In particular, I want to thank Donald Blackmore Wagner (Berkeley) who put at my disposal his English translation of the most interesting parts of the Chinese "Nine Chapters of the Art of Arithmetic" and of Liu Hui's commentary to this classic, and also Jacques Sesiano (Geneva), who kindly allowed me to use his translation of the recently discovered Arabic text of four books of Diophantos not extant in Greek. Warm thanks are also due to Wyllis Bandler (Colchester, England) who read my English text very carefully and suggested several improvements, and to Annemarie Fellmann (Frankfurt) and Erwin Neuenschwander (Zürich) who helped me in correcting the proof sheets. Miss Fellmann also typed the manuscript and drew the figures.

I also want to thank the editorial staff and production department of Springer-Verlag for their nice cooperation.

Zürich, February, 1983   B. L. van der Waerden

Table of Contents

Introduction

Until quite recently, we all thought that the history of mathematics begins with Babylonian and Egyptian arithmetic, algebra, and geometry. However, three recent discoveries have changed the picture entirely.

The first of these discoveries was made by A. Seidenberg. He studied the altar constructions in the Indian Śulvasūtras and found that in these relatively ancient texts the "Theorem of Pythagoras" was used to construct a square equal in area to a given rectangle, and that this construction is just that of Euclid. From this and other facts he concluded that Babylonian algebra and geometry and Greek "geometrical algebra" and Hindu geometry are all derived from a common origin, in which altar constructions and the "Theorem of Pythagoras" played a central rôle.

Secondly I have compared the ancient Chinese collection "Nine Chapters of the Arithmetical Art" with Babylonian collections of mathematical problems and found so many similarities that the conclusion of a common pre-Babylonian source seemed unavoidable. In this source, the "Theorem of Pythagoras" must have played a central rôle as well.

The third discovery was made by A. Thom and A. S. Thom, who found that in the construction of megalithic monuments in Southern England and Scotland "Pythagorean Triangles" have been used, that is, right-angled triangles whose sides are integral multiples of a fundamental unit of length. It is well-known that a list of "Pythagorean Triples" like (3,4,5) is found in an ancient Babylonian text, and the Greek and Hindu and Chinese mathematicians also knew how to find such triples.

Combining these three discoveries, I have ventured a tentative reconstruction of a mathematical science which must have existed in the Neolithic Age, say between 3000 and 2500 B.C., and spread from Central Europe to Great Britain, to the Near East, to India, and to China. By far the best account of this mathematical science is found in Chinese texts. My ideas concerning this ancient science will be explained in Chapters 1 and 2.

The Greeks had some knowledge of this ancient science, but they transformed it completely, creating a deductive science based on definitions, postulates and axioms. Yet several traces of pre-Babylonian geometry and algebra can be discerned in the work of Euclid and Diophantos and in popular Greek mathematics. This will be shown in Chapters 3, 4 and 6.

In the treatises of Hindu astronomers like Āryabhaṭa and Brahmagupta, who lived in the sixth and seventh century A.D., we find methods to solve Diophantine equations such as

$$ax + c = by$$

and

$$x^2 = Dy^2 + 1.$$

These methods are based on the Euclidean algorithm. In Chapter 5 I shall give an account of these methods and discuss their relation to Greek science.

Chapter 7 deals with the work of the excellent Chinese geometer Liu Hui (third century A. D.) and with some mathematical passages in the work of the great Indian astronomer Āryabhaṭa (sixth century). It seems to me that both were influenced by the work of Greek geometers and astronomers like Archimedes and Apollonios. In particular I shall discuss Liu Hui's measurement of the circle and Āryabhaṭa's trigonometry.

Chapter 1

Pythagorean Triangles

Part A

Written Sources

Fundamental Notions

A *Pythagorean Triangle* is a right-angled triangle in which the three sides are proportional to integers x, y, and z. According to the "Theorem of Pythagoras", the integers must satisfy the equation

$$(1) \qquad x^2 + y^2 = z^2.$$

A *Pythagorean Triple* is a triple of integers (x,y,z) satisfying (1). The triple is *primitive* if x,y,z have no common factor.

In a primitive Pythagorean triple one of the numbers x and y must be odd and the other even, because if both are even, the triple is not primitive, and if both are odd, the sum $x^2 + y^2$ is a number of the form $4n+2$ and cannot be a square.

The construction of primitive Pythagorean triples is a simple arithmetical problem. One can start with any integer x and solve the equation

$$(2) \qquad (z-y)(z+y) = x^2$$

for y and z. If one starts with an odd number $x = st$, one can satisfy the equation (2) by taking

$$(3) \qquad \begin{array}{cc} z+y=s^2 & z = \tfrac{1}{2}(s^2+t^2) \\ z-y=t^2 & y = \tfrac{1}{2}(s^2-t^2). \end{array}$$

As we shall see presently, this method of solution was used in a Chinese text from the Han-period (about -200 to $+220$). On the other hand, if one starts with an even integer $x = 2pq$, one can satisfy (2) by taking

(4)
$$z + y = 2p^2 \quad z = p^2 + q^2$$
$$z - y = 2q^2 \quad y = p^2 - q^2.$$

This solution was used by Diophantos of Alexandria. In book VI of his Arithmetica the first problem reads:

Problem 1. To find a right-angled triangle such that the hypotenuse minus each of the other sides gives a cube.

The solution begins thus:

Let the triangle be formed with the aid of two numbers, say s and 3. The hypotenuse is then $s^2 + 9$, the height $6s$ and the base $s^2 - 9$.

In our text-books of elementary number theory it is proved that all primitive Pythagorean triples can be obtained from the formulae (4). Just so, one can prove that all primitive Pythagorean triples can be obtained from (3). In fact, (4) can be obtained from (3) by substituting

$$s = p + q, \quad t = p - q$$

and interchanging x and y. So the two methods for finding Pythagorean triples, the Chinese and the Greek method, are equivalent.

The Text Plimpton 322

Long before Diophantos, the Babylonians already knew how to calculate Pythagorean triples. To see this, let us consider the cuneiform text "Plimpton 322", written under the dynasty of Hammurabi, and published by Neugebauer and Sachs in their book "Mathematical Cuneiform Texts" (New Haven 1945). It has been discussed by Peter Huber (l'Enseignement mathématique 3, p. 19), by Derek de Solla Price (Centaurus 10, p. 219–231), and by E. M. Bruins (Physis 9, p. 373–392). Here I shall use only those facts and interpretations on which all authors agree.

The text is the right-hand part of a larger tablet containing several columns of numbers. The last column contains nothing but the numbers $1, 2, \ldots, 15$. The preceding columns refer, according to the legend at the head of the columns, to the "width" and the "diagonal" (of a rectangle). I shall denote the "width" by y and the "diagonal" by z. If one calculates $z^2 - y^2$, one finds that it is in all the cases the square of an integer x containing only factors 2, 3 and 5. Such "regular numbers" x play a special role in the sexagesimal number system, because the reciprocals x^{-1} can be written as finite sexagesimal fractions. The Babylonians had tables of reciprocals of these "regular numbers".

A few errors in the text have been corrected already by Neugebauer and Sachs. After these corrections, the 15 triples (x, y, z) are:

x	y	z
2, 0	1,59	2,49
57,36	56, 7	1,20,25
1,20, 0	1,16,41	1,50,49
3,45, 0	3,31,49	5, 9, 1
1,12	1, 5	1,37
6, 0	5,19	8, 1
45, 0	38,11	59, 1
16, 0	13,19	20,49
10, 0	8, 1	12,49
1,48, 0	1,22,41	2,16, 1
1, 0	45	1,15
40, 0	27,59	48,49
4, 0	2,41	4,49
45, 0	29,31	53,49
1,30	56	1,46

The numbers are written in the sexagesimal system based on powers of 60. For instance, the number 2,49 in the first line means

$$2 \times 60 + 49 = 169 .$$

In the text, the two columns headed y and z are preceded by a column representing either

$$A = (y/x)^2$$

or

$$1 + A = (x^2 + y^2)/x^2 = (z/x)^2 .$$

For instance, in the first line, we have

$$x = 2, 0 \quad \text{and} \quad y = 1,59$$

hence

$$y/x = 0;59,30$$

which means $59/60 + 30/60^2$. We now have

$$A = (y/x)^2 = 0;59, 0,15$$

and

$$1 + A \qquad = 1;59, 0,15 .$$

We cannot distinguish between the two possibilities A and $1+A$, because in the first column the initial digits 1, if they ever existed, are broken off.

All commentators agree that the preserved columns were probably preceded by at least three columns, in which the quotients

$$y/x=v \quad \text{and} \quad z/x=w$$

and the height x were recorded. This means: the Babylonians first calculated a Pythagorean triple $(1,v,w)$ satisfying the equation

(5) $$1+v^2=w^2$$

and next multiplied the triple $(1,v,w)$ by a suitable number x in order to obtain integer triples (x,y,z).

There are three arguments in favour of this hypothesis:

First, in one case (line 11 of the tablet) the triple $(1,v,w)$ would be

$$1, \quad v=0;45, \quad w=1;15.$$

A multiplication by 4 would have yielded the well-known Pythagorean triple $(4,3,5)$. Instead of performing this easy multiplication, the scribe left the triple $(1,v,w)$ as it stands and wrote

$$1, \quad 45, \quad 1,15.$$

Note that in the Babylonian notation we cannot distinguish between $1,15=75$ and $1;15=1\ 15/60$.

The *second* argument is: the Babylonians had a column

$$A=v^2 \quad (\text{or} \ 1+A=w^2).$$

This column was probably derived from an earlier column v (or w) by squaring.

The *third* argument is: it is easier to solve the equation (5) than the equation (1), for (5) can be written as

(6) $$(w+v)(w-v)=1.$$

In all 15 cases, $w+v$ is a regular sexagesimal number. If this number is called d, we have

$$w+v=d$$
$$w-v=d^{-1}$$

from which v and w can be solved easily.

It is possible that the Babylonians took d and d^{-1} directly from a table of reciprocals. It is also possible that they put

$$d=p/q=pq^{-1}$$

and

$$d^{-1}=q/p=qp^{-1}$$

as Neugebauer and Sachs supposed. I cannot decide between these possibilities.

If the Pythagorean triples of Plimpton 322 were computed by this method, the Babylonian scribes must have known the "Theorem of Pythagoras"
as well as the algebraic identity

(7) $$z^2 - y^2 = (z+y)(z-y).$$

Both conclusions are confirmed by other Babylonian texts. For instance, in the text BM 85196 the hypotenuse $z = 30$ and the height $y = 24$ of
a right-angled triangle are given, and the base $x = 18$ is calculated as the
square root of $z^2 - y^2$ (see my Science Awakening I, p. 76). The identity (7)
was used in many Babylonian problem solutions, as we shall see in Chapter 2. Note: the signature BM means British Museum.

All in all, we may safely conclude that the Babylonians calculated Pythagorean triples (x,y,z) by first assuming x to be 1, and next multiplying
the triple $(1,v,w)$ by a suitable integer.

Can we find similar methods in other civilizations? Yes, we can.

A Chinese Method

The Chinese collection of mathematical problems "Nine Chapters on
the Mathematical Art"[1], written during the Han-period (circa -200 to
$+220$), contains in Chapter 9 a sequence of problems on right-angled triangles. The problems will be discussed in greater detail in Chapter 2. For the
moment, I only note that all solutions are based on the "Theorem of Pythagoras". For instance, in Problem 1, x and y are given, and z is computed as
the square root of $x^2 + y^2$.

In the 16 problems of Chapter 9, the following Pythagorean triangles
occur:

$$
\begin{array}{ccc}
3 & 4 & 5 \\
5 & 12 & 13 \\
8 & 15 & 17 \\
7 & 24 & 25 \\
20 & 21 & 29.
\end{array}
$$

How did the author of the problems find these triangles? Light upon
this question is shed by the very remarkable problem 14. The following
translation (with explanations in parentheses) is due to Donald Blackmore
Wagner, who kindly put at my disposal his translations of several problems
of Chapter 9. Problem 14 reads:

---

1 German translation by Kurt Vogel: Neun Bücher arithmetischer Technik (Vieweg,
  Braunschweig 1968).

Two persons are standing together. The proportion of A's walking is 7, and the proportion of B's walking is 3. (That is, the ratio of their walking speeds is $a:b=7:3$).

B walks east. A walks south $10\,pu$, then walks diagonally (roughly) northeast, and meets B. How far do A and B walk?

Anwer: B walks $10\frac{1}{2}\,pu$. A walks diagonally $14\frac{1}{2}\,pu$.

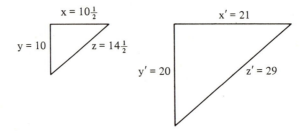

Fig. 1. A Chinese walking problem

To make things clearer, I have drawn two similar rightangled triangles $(10,10\frac{1}{2},14\frac{1}{2})$ and $(20,21,29)$. The Chinese text has no drawings. The text explains the method of solution thus:

Method: Multiply 7 by itself; also multiply 3 by itself. Add (the products) and halve. This is the proportion of A walking diagonally ($z'=29$).

Subtract from the proportion (of A walking diagonally) the product of 3 by itself. The difference is the proportion of walking south ($y'=29-9=20$).

Let the product of 3 and 7 be the proportion of B walking east ($x'=21$).

Set up the $10\,pu$ walked south, and multiply by the proportion of A walking diagonally (result $yz'=10\times29=290$). Again set up $10\,pu$ and multiply by the proportion of B walking east (result $yx'=10\times21=210$). Let each (result) be a dividend. Dividing the dividends by the proportion of walking south ($y'=20$) gives in each case the number (of pu) walked.

In the statement of the problem as well as in the solution, a mathematical terminus is used which Wagner has translated as "proportion". The expression "the proportion of A's walking is 7, and the proportion of B's walking is 3" means: the distances traversed by A and B in the same time are in the ratio of 7 to 3. In Fig. 1, this means

(8) $$z+y=(7/3)x.$$

We also are given

(9) $$y=10.$$

The method by which the Chinese author calculates the sides x, y, z of the required triangle is very remarkable. He first finds a Pythagorean triple (x', y', z') satisfying the condition (8). He calls x' "the proportion of B walking east", and similarly for y' and z'. That is, the author knew that he had to make x, y, z proportional to x', y', z' in order to reach his aim. The latter triple was calculated as follows, starting with the given numbers 7 and 3:

$$7 \times 7 = 49$$
$$3 \times 3 = 9$$
$$z' = \tfrac{1}{2}(49 + 9) = 29$$
$$y' = 29 - 9 = 20$$
$$x' = 3 \times 7 = 21$$

We now have

$$z' - y' = 9 \ = 3 \times 3$$
$$z' + y' = 49 = 7 \times 7$$
$$x' = 21 = 3 \times 7.$$

It follows that x', y', z' satisfy conditions (2) and (8). Next these three numbers are multiplied by $y/y' = 10/20$ in order to satisfy the condition (9).

In all problems of Chapter 9, the data are presented as special numbers such as 7 and 3, but the methods of solution are formulated as general rules, which can adequately be represented by modern formulae such as

$$x' = a \cdot b$$
$$z' = \tfrac{1}{2}(a^2 + b^2).$$

The second side, the "proportion of A walking south", is calculated as

$$y' = z' - b^2$$

but this is equivalent to

$$y' = \tfrac{1}{2}(a^2 - b^2).$$

So the Chinese rule for computing Pythagorean triples can be written as

(10)
$$\begin{cases} x' = ab \\ y' = \tfrac{1}{2}(a^2 - b^2) \\ z' = \tfrac{1}{2}(a^2 + b^2). \end{cases}$$

As we have seen, all primitive Pythagorean triples in which x is odd, can be obtained by these formulae. So the Chinese method is perfectly general.

Methods Ascribed to Pythagoras and Plato

In Proklos' commentary to the first book of Euclid's Elements, on p. 340 in the translation of Morrow (p. 428 in the edition of Friedlein), a method for obtaining Pythagorean triples is ascribed to Pythagoras. One takes an odd number a and puts

(11)
$$\begin{cases} x = a \\ y = \frac{1}{2}(a^2 - 1) \\ z = \frac{1}{2}(a^2 + 1). \end{cases}$$

It follows that $z - y = 1$ and $z + y = a^2$, so the equation

$$(z + y)(z - y) = x^2$$

is satisfied. All triples in which $z - y = 1$ can be obtained from (11). Obviously, (11) is a special case of (10), so the "Method of Pythagoras" is closely related to the Chinese method.

Another method is ascribed to Plato. I quote from Morrow's translation:

> The Platonic method proceeds from even numbers. It takes a given number as one of the sides about the right angle, divides it into two and squares the half, then by adding one to the square gets the subtending side, and by subtracting one from the square gets the other side about the right angle.

The resulting triples can be written as

(12)
$$\begin{cases} x = 2c \\ y = c^2 - 1 \\ z = c^2 + 1. \end{cases}$$

In this case we have $z - y = 2$ and $z + y = 2c^2$. All triples with $z - y = 2$ can be obtained by this method.

Pythagorean Triples in India

The Śulvasūtras are ancient Hindu manuals in which detailed prescriptions for the construction of altars of given form and size are given. The word Śulvasūtra means something like "Manual of the Cord", and in fact the instruments prescribed for the altar constructions are peg and cord.

The Śulvasūtras were probably composed between 500 and 200 B.C. For the dating problem see L. Renou and J. Filliozat: L'Inde Classique (Paris 1947), p. 302. Some geometrical constructions used in the Śulvasūtras are already mentioned in the earlier Śatapatha Brāhmaṇa, as we shall see later. According to Renou and Filliozat (L'Inde Classique, p. 267), this Brāhmaṇa was written between 1000 and 800 B.C.

A recent study of the geometrical methods used in the Śulvasūtras is the book of Axel Michaels: Beweisverfahren in der vedischen Sakralgeometrie (Franz Steiner, Wiesbaden 1978).

For our present purpose most interesting is the Śulvasūtra ascribed to Baudhāyana. It has been published and translated by G. Thibaut in The Pandit 9 (1874), 10 (1875) and New Series 1 (1876–77). It contains a passage on the Theorem of Pythagoras and, immediately following, a passage on Pythagorean triangles. The two passages read:

The diagonal of an oblong produces by itself both the areas which the two sides of the oblong produce separately (i.e. the square of the diagonal is equal to the sum of the squares of the two sides).

This is seen in those oblongs the sides of which are three and four, twelve and five, fifteen and eight, seven and twenty-four, twelve and thirty-five, fifteen and thirty-six.

The Pythagorean triples mentioned in the second passage can be classified thus:

$z-y=1$	$z-y=2$	$z-y=3$
3, 4, 5	8, 15, 17	15, 36, 39
5, 12, 13	12, 35, 37	
7, 24, 25		

It is readily seen that the triples in the first two columns can be obtained from the rules ascribed to Pythagoras and Plato. The triple in the third column is just 3 times (5,12,13). In all these triples $z-y$ is small, which makes the calculation of the triples by means of the formula

(13) $$(z+y)(z-y)=x^2$$

very easy. The triple (20,21,29), which is not so easy to obtain, is not mentioned by Baudhāyana.

One might ask whether Baudhāyana or one of his predecessors found these triples empirically by measuring triangles. I don't think that this is probable, for several reasons.

First, there are more than 1 000 triangles with integer sides x and y and $x+y<48$. In order to find the triple (12,35,37) empirically, one would have to measure the hypotenuses of a considerable portion of these triangles.

Secondly, the author of the Baudhāyana Śulvasūtra knew the equation

(14) $$z^2-y^2=x^2$$

and used it in his altar constructions. It is very easy to rewrite this equation in the equivalent form (13) and to solve it for small values of $z-y$ by a suitable choice of x.

Thirdly, long before Baudhāyana the Babylonians already computed Pythagorean triples, starting with the Eq. (13) as we have seen.

All in all, it seems most likely that the triples were calculated by means of the Eq. (13).

The Hypothesis of a Common Origin

We have seen that methods for computing Pythagorean triples were practiced at an early date in *Babylonia,* in *India,* in *Greece,* and in *China.* We have also seen that the Theorem of Pythagoras was known in all four countries, and that the methods by which the triples were computed were closely related: they all started, most probably, from the Eq. (10).

If one knows the Theorem of Pythagoras, it is possible, but not at all necessary to compute Pythagorean triples. Our physicists and engineers often use the Theorem of Pythagoras, but they have no use for Pythagorean triples. In the work of Euclid, Archimedes, and Apollonios no Pythagorean triples are found. So, if we find Pythagorean triples associated with the Theorem of Pythagoras in four ancient civilizations, we may conclude that *a common origin of the whole theory is highly probable.*

With very few exceptions, great discoveries in mathematics, physics, and astronomy have been made only once. Epicycles and eccenters, the spherical form of the earth, the heliocentric system, the three laws of Kepler, the three laws of Newton's mechanics, the law of gravitation, they all were discovered only once. The same holds for the laws of optics, of electricity and magnetism, and so on. In mathematics there are a few cases of independent invention, for instance the discovery of non-Euclidean geometry by Gauss, Bolyai and Lobatchevski, but the overwhelming majority of great discoveries in geometry, algebra and analysis were made only once. Therefore, when we find that a great and important theorem, like that of Pythagoras, which is by no means easy to find, is known in several countries, *the best thing to do is, to adopt the hypothesis of dependence.*

It seems that many historians don't like this heuristic principle. They very often assume independent invention, which means that they don't even try to explain similarities. I, for my part, have found that the hypothesis of dependence is a very useful working hypothesis. If one starts with this hypothesis, one often finds "missing links" or other indications of connections and interactions between civilizations.

These general considerations are confirmed in the case of the Theorem of Pythagoras. We shall see that there are many more points of contact between Greek and Hindu mathematics, and also between Babylonian and Chinese mathematics. In the present chapter I shall show that some quite specific geometrical constructions are found in the Elements of Euclid as well as in the Brāhmaṇas and Śulvasūtras. Also, some ideas about the ritual importance of exact geometrical constructions are found in Greece as well as in India.

Geometry and Ritual in Greece and India

In a pioneer paper The Ritual Origin of Geometry, Archive for History of Exact Sciences 1 (1963), and again in a later paper The Origin of Mathematics in the same Archive 18 (1978), A. Seidenberg has pointed out that in

Greek texts as well as in the Śulvasūtras geometrical constructions were regarded important for ritual purposes, namely for constructing altars of
given form and magnitude. In Greece this led to the famous problem of
"doubling the cube", whereas in India it was not the volume but the area
of the altar that was considered important. In both cases, one essential step
in the altar constructions was the solution of the problem: *to construct a
square equal in area to a given rectangle.* To solve this problem, exactly the
same construction was used in Greece and in India, a solution based on the
Theorem of Pythagoras. Also, the ideas about the religious importance of
exact geometrical altar constructions were very similar in both countries.
From these facts Seidenberg concluded that a common origin of these
geometrical and religious ideas must be assumed.

I shall now explain Seidenberg's ideas in greater detail.

A fundamental problem in Greek geometry is: to construct a mean proportional line segment x between two given line segments a and b:

$$a:x=x:b.$$

As Aristotle points out, this problem is solved as soon as one can construct a square equal in area to the rectangle spanned by a and b:

$$x^2=ab.$$

In the second book of Euclid's Elements, proposition 14, the latter
problem is solved in several steps. First, by taking away from the rectangle
with base a and height b a small rectangle $(a-b)/2 \cdot b$ (the shaded rectangle
on the right in Fig. 2) and adding an equal rectangle at the top, one obtains
a difference of two squares. The result may be expressed by the modern
formula

$$a \cdot b = \left(\frac{a+b}{2}\right)^2 - \left(\frac{a-b}{2}\right)^2.$$

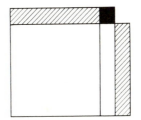

Fig. 2. How to transform a rectangle into a difference of two squares

Next, by an application of the Theorem of Pythagoras, a difference of
two squares can be made equal to a square

$$z^2 - y^2 = x^2.$$

Thus, in two steps, Euclid obtains a square x^2 equal to the given rectangle $a \cdot b$.

In the Śulvasūtras, problems of the following kind are treated: Given an altar in falcon-shape (see Fig. 3) having an area of $7\frac{1}{2}$ *purushas,* to construct an altar of exactly the same shape having an area of $8\frac{1}{2}$ *purushas.* In the course of the construction, the problem arises to construct a square equal in area to a given rectangle. This problem is solved by just the same two steps as in Euclid's Elements. In a first step the rectangle is transformed into a difference of two squares, and next this difference is made equal to a square by means of the Theorem of Pythagoras. Seidenberg has shown that this construction was known already to the author of the Satapatha Brāhmaṇa. This author lived, in all probability, before 600 B.C., so he cannot have been influenced by Greek geometry.

In Euclid's book 2, no proportions are used, only additions and subtractions of plane areas. However, if one uses proportions, there are several

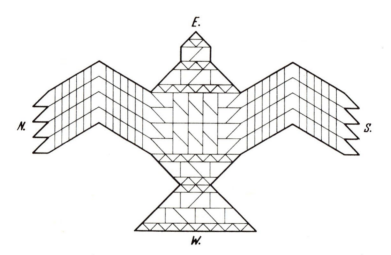

Fig. 3. Altar in Falcon-Shape

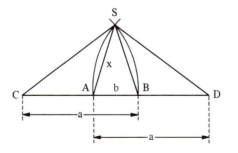

Fig. 4. Construction of a mean proportional

other methods to construct a mean proportional x between two given line segments a and b. For instance, if one makes

$$AB=b$$

and

$$AD=BC=a$$

(see Fig. 4), and if S is one of the points of intersection of two circles with radius a centered at C and D, the isosceles triangles SAB and DAS are similar, and the line $AS=x$ is a mean proportional between a and b.

This is only one of several possible constructions. The fact that Euclid and the Śulvasūtras both use one and the same more complicated construction, based on the Theorem of Pythagoras, is a strong argument in favour of a common origin.

On p. 322 of his paper The Origin of Mathematics (Archive 18), Seidenberg adduces another, still stronger argument. After having sketched the construction of the altar of $7\frac{1}{2}$ *purushas* and its augmentation of one *purusha* at the time while maintaining similarity of form, the Śatapatha Brāhmaṇa says:

He (the sacrificer) thus expands (the wing) by as much as he contracts it, and thus, indeed, he neither exceeds (its proper size) or does he make it too small

and next:

Those who deprive the agni (altar) of its true proportions will suffer the worse for sacrificing.

Now the same idea, namely that a suffering sent by the gods can only be avoided by performing an exact geometrical construction, also occurs in Greek texts. In the dialogue "Platonikos" of Eratosthenes a story was told about the problem of doubling the cube[2]. According to this story, as Theon of Smyrna recounts it in his book "Exposition of mathematical things useful for the reading of Plato", the Delians asked for an oracle in order to be liberated from a plague. The god (Apollo) answered through the oracle that they had to construct an altar twice as large as the existing one without changing its shape. The Delians sent a delegation to Plato, who referred them to the mathematicians Eudoxos and Helikon of Kyzikos.

It is very remarkable that in Greece as well as in India, ritual requirements regarding the shape and volume or area of an altar were combined with the idea of sufferings sent by the gods, which one might avert by performing exact geometrical constructions.

Greek and Sanskrit are both Indo-European languages. All Indo-European languages are derived from an original language or group of related dialects. Now the spread of languages is in many cases connected with the spread of civilizations and religions. Examples: the spread of the Greek language in the Hellenistic age, of Latin in the Roman Empire, of Spanish and Portuguese to South America, of English to North America and Australia. Thus, it is only natural to suppose that the spread of Indo-European

---

2 For the proof that "Platonikos" was a dialogue see P. Wolfer: Eratosthenes als Mathematiker und Philosoph, Dissertation Zürich 1954.

languages in the third and second millennium B.C. was accompanied by
the spread of ritual and mathematical ideas.

The Greek and Hindu constructions of a square equal to a given rectangle are both based on the Theorem of Pythagoras. The Greek and Hindu
computations of Pythagorean triples and triangles are also based on this
theorem. So it seems that the Theorem of Pythagoras was known already
about 2000 B.C., when the ancestors of the Greeks and the Hindu Aryans
were still close together in the Danube region. From this region, the Greeks
went southward, the Aryans south-eastward to Iran and India.

Pythagoras and the Ox

There is still another legend indicating that altar offerings and geometrical theorems were somehow connected in the mind of the ancient
Greeks. The legend is well known, but it seems that before Seidenberg no-
body has seen its significance.

In Euclid's Elements, the Theorem of Pythagoras appears as Proposition I,47. In his commentary to this proposition, Proclos writes:

If we listen to those who wish to recount ancient history, we may find some who refer this
theorem to Pythagoras, and say that he sacrificed an ox in honour of the discovery.

And Plutarch, who lived in the first century A.D., quotes a distich:
"When Pythagoras discovered his famous figure, for which he sacrificed a
bull, ..."

If this saying about Pythagoras is considered as a statement of fact, it is
extremely improbable, for Pythagoras was strongly opposed to the sacrifice
of animals, especially cattle, and this attitude was well known among later
authors (see e.g. my book "Die Pythagoreer", p. 17–19). But if the saying is
considered just as a legend, it is easy to explain how such a legend might
have come into existence. I suppose that the original form of the legend
said something like "He who discovered the famous figure sacrificed a bull
in honour of the discovery." I suppose the discovery was made before the
time of Hammurabi, for at this time the theorem was known and incorpo-
rated into Babylonian problem texts. I also suppose that the discoverer was
a priest entitled to sacrifice animals, and at the same time a mathematician.

Legends are often transferred to other famous persons. In this case, it
was only natural that "he who discovered the famous figure" (or the famous theorem) was identified with Pythagoras.

I suppose that the man who discovered the theorem and applied it to
altar constructions belonged to a community of priests comparable with
the Hindu priests who composed the Brāhmaṇas and Śulvasūtras.

Part B

Archaeological Evidence

Up to this point, we have used only written sources, but we cannot stop here. If it is true that the mathematical ideas found in Babylonian, Indian, Greek, and Chinese sources have a common origin, these ideas must have originated at a time when writing was still unknown. In order to find out more about this common origin, we have to dive into prehistory and to use archaeological evidence.

Prehistoric Ages

The Stone Age, the age of stone implements, is usually divided into three periods, namely
the *Palaeolithic Age,* the time of the marvellous cave paintings in Spain and southern France,
the *Mesolithic Age,* an intermediate period in which civilization was at a low level, as compared with the preceding and following periods,
the *Neolithic Age,* characterized by extremely fine stone implements and beautiful pottery, and by the development of agriculture, cattle-breeding, shipbuilding, navigation, and megalithic architecture.
In Western and Central Europe, the Neolithic Age lasted from about 4500 B.C. to about 2000 B.C. The last part of this period, the *Chalcolithic Age,* beginning about 2500 B.C., is characterized by the appearance of copper. After 2000 B.C., the *Bronze Age* begins.

Radiocarbon Dating

All approximate dates presented in this chapter are *tree-ring corrected radiocarbon dates.* Let me explain what this means.
The method of radiocarbon dating, introduced by W. F. Libby in 1946, is based on two facts and two hypotheses. The facts are:
F1. Every plant assimilates, as long as it lives, carbon atoms from the atmosphere. Therefore the ratio of radioactive C-14 to normal C-12 in a living plant or animal is practically the same as in the atmosphere.
F2. After death the amount of C-14 in a plant or animal decreases according to the well-known law of radioactive disintegration.
Libby's hypotheses were:
H1. The half-life of C-14 is 5568 years.
H2. The ratio of C-14 to C-12 in the atmosphere has been constant from prehistoric times until the first atomic bomb explosion.
Under these two assumptions the date of any organic material found in an excavation can be estimated by measuring its radioactivity.

Unfortunately, Libby's two assumptions H1 and H2 are only approximately true. The half-life of C-14 is about 3% larger than Libby's estimate, and the concentration of C-14 in the atmosphere is variable. As a consequence, the true dates are in most cases earlier than the dates found by Libby's method. In the third millennium B.C., which is the most interesting period in our present investigation, the difference may amount to several centuries.

A good method for correcting pre-historic radiocarbon dates is the *tree-ring method*. By comparing tree-ring sequences of long-living trees, a tree-ring chronology has been established, which can be used for correcting radiocarbon dates. Correction tables were set up by Suess and other authors. See for this subject the article of H. MacKerrell "Correction Procedures for C-14 Dates" in: Radiocarbon, Calibration and Prehistory, edited by T. Watkins, Edinburgh 1975.

Tree-ring corrected radiocarbon dates are not completely reliable. First, they have random errors due to radiation counting, which may amount to 50 or 100 years. Secondly, a comparison of radiocarbon dates with Egyptian historical dates seems to show that the tree-ring corrected radiocarbon dates may have systematic errors amounting to 100 or 200 years for the Neolithic Age.

Megalithic Monuments in Western Europe

The investigations of A. Thom and A.S. Thom, of G. Hawkins, F. Hoyle and others on the geometry of megalithic monuments in Brittany (Bretagne), England, Scotland, and Ireland have been summarized and critically commented upon by J.E. Wood in his book Sun, Moon and Standing Stones, Oxford 1978. I shall use this book as my main source of information[3]. For a general view of the way of living of the megalith builders and their science, the reader may consult two highly interesting books of E. MacKie:

Science and Society in Prehistoric Britain, London 1977,

The Megalith Builders, Oxford 1977.

The latter book also takes into account megalithic monuments in Portugal, Spain, and Malta.

The earliest megalithic monuments are "passage graves" in Portugal, in Spain and in Brittany, built between 4800 and 3000 B.C. A passage grave consists basically of a round burial chamber in the middle of a mound and a stone passage leading to the chamber. Sometimes side chambers were added.

The monuments are never far from the coast, and are of nearly the same type in all these countries. The skeletons found in the graves show that the people who were buried here were long-headed and of short stature, and

---

3 See also A. Thom and A.S. Thom: Megalithic Remains in Britain and Brittany, Oxford 1978, and other books and papers of the Thoms.

racially rather homogeneous, of a type that occurs most frequently in Mediterranean countries. They seem to have formed a ruling class that came by sea from Portugal to Brittany, and next to England, Scotland and Ireland. They must have had considerable skill in shipbuilding, navigation and architecture.

About 3300 B.C., the beautiful passage grave of New Grange in Ireland was constructed. Its entrance was built in such a way that on the day of the winter solstice the rays of the rising sun, shining through the "roof box" and the covered passage, just reached the burial chamber.

This is not the only case of astronomical orientation. Two passage graves at Clava in Scotland were oriented to the midwinter sunset. Obviously, the people who built these graves considered the summer and winter solstices as important events.

In the centuries just before and after 3000 B.C. the same people began to build "henges", that is meeting places within earthen enclosures. The most famous henge, Stonehenge in Wiltshire, was oriented towards the midsummer sunrise. From the center of the circular enclosure one could see the sun rise over the "Heelstone" on the day of the summer solstice. The enclosing circular bank and ditch and the heelstone belong to Stonehenge I, which was built about 2800 B.C. Stonehenge II and III, the most impressive parts of the monument, are later additions.

Most "henges" are in Southern England. As a rule, they were not inhabited: they seem to have been mainly ceremonial meeting places, and probably also astronomical observatories. However, at Durrington Walls in Wiltshire, the excavations have brought to light ash and other refuse, a spindle whorl and a large amount of animal bones, all pointing towards a permanent habitation. Among eight thousand or more animal bones there were practically no skulls. From this fact, MacKie concluded that meat was brought to the site from elsewhere, and that the inhabitants of Durrington Walls were not peasants or hunters, but some kind of élite group, supported by gifts or tributes of food from the local rural population. From the finds in the stone village of Skara Brae on the Orkney islands MacKie was able to draw an similar conclusion.

Soon after 2500 B.C. the "Beaker People", so called after their typical pottery ("Glockenbecher") came to England. Skeletons found in England and Central Europe show that the Beaker People were tall and had characteristic round heads. In England they mixed with the earlier, long-headed ruling class, and they continued to use the megalithic henges. The Beaker People erected the marvellous trilithons of Stonehenge II and III. Stonehenge II ist late neolithic (about 2100), and Stonehenge III was built in the Bronze Age.

Pythagorean Triples in Megalithic Monuments

In many henges we find rings of standing stones. Most rings, including the earliest ones, are true circles, but some are accurately constructed ellipses. It is well-known that an ellipse can be constructed by spanning a rope around two pegs located at the foci F_1 and F_2 (see Fig. 5). The length of the rope is equal to the distance $F_1 F_2$ plus the major axis of the ellipse. In the triangle $O B F_1$ the side a is equal to one-half of the major axis, b is one-half of the minor axis, and c is one-half of the distance $F_1 F_2$.

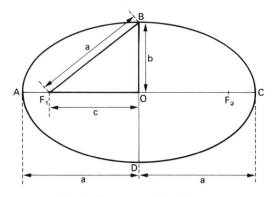

Fig. 5. Geometry of the Ellipse

In some henges, the triangle $O B F_1$ is a Pythagorean triangle. One of the ellipses at Callanish on the Hebrides is based on a (3,4,5) triangle. Two ellipses at Stanton Drew in Somerset have (5,12,13) triangles, and an ellipse at Daviot near Inverness in Scotland has a (12,35,37) triangle.

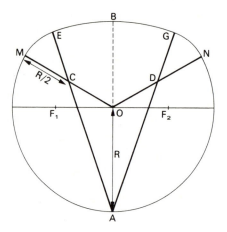

Fig. 6. Type A flattened circle

Other rings are composed of arcs of circles. We find "flattened circles" of two types A and B (see Fig. 6 and Fig. 7). Type B, the simpler of the two, is made up of a semicircle with radius OA, two arcs of circles with radius

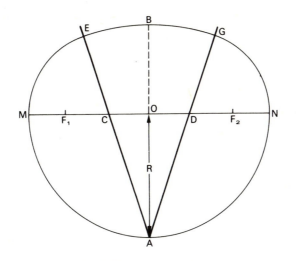

Fig. 7. Type B flattened circle

$CM = DM$ centred on C and D, and a joining arc EG centred on A. More than a dozen stone rings of this type are known. – In Type A the centres C and D are not on a diameter of the basic circle, but are in the mid-points of two radii OM and ON with an angle of $120°$ between them.

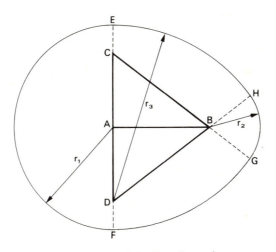

Fig. 8. Geometry of the Type I egg-shape

Other rings are egg-shaped. The Type I egg (Fig. 8) has two triangles placed back to back. Its perimeter consists of the arcs of four circles, one centred at A, two centred at C and D with radii $DE = CF$, and one centred at B with radius BH. The (3,4,5) triangle was often used in the construction of the eggs, but at Woodhenge the triangle was (12,35,37), if Thom's inter-

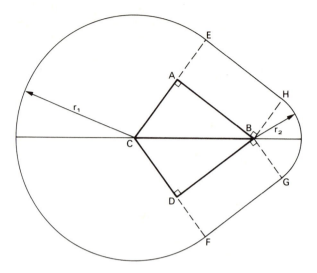

Fig. 9. Geometry of the Type II egg-shape

pretation is correct. Our Fig. 10 is based on a diagram in the book of A. and A.S. Thom mentioned before. The Figs. 5–9 illustrating the construction of the stone rings are all taken from the book of J.E. Wood: Sun, Moon and Standing Stones.

Woodhenge was, most probably, the site of a large roofed wooden building. The concentric Type I eggs of Woodhenge have their centres at the vertices of two congruent triangles ABC and ABD. The sides of the triangle ABC are half-integer multiples of the "Megalithic Yard" (MY) of 0.829 m, namely

$$AB = 6\,MY, \qquad AC = 17\tfrac{1}{2}\,MY, \qquad BC = 18\tfrac{1}{2}\,MY.$$

If these numbers are doubled, the Pythagorean triple (12,35,37) is obtained. The same triple was also found at Daviot near Inverness, as we have seen. The Megalithic Yard seems to have been a fundamental unit in many megalithic constructions.

Another example of a Type I egg based on a (3,4,5) triangle, the sides of which are simple multiples of the Megalithic Yard, is a stone ring at Clava in Inverness, which surrounds a circular passage grave (Fig. 11).

Type II eggs are similar to Type I eggs, but the triangles are now placed with their hypotenuses together (Fig. 9). Two rings of this type, namely The Hunters on Bodmin Moor in Cornwall and Buckland Ford on Dartmoor, are based on (3,4,5) triangles.

The question now arises: Are Thom's interpretations completely certain? From Figures 6 and 7 one sees that the stones and holes are not always exactly on the circles. The positions of the points B, C and D in Fig. 10 can be varied a little without spoiling the goodness of fit. In the

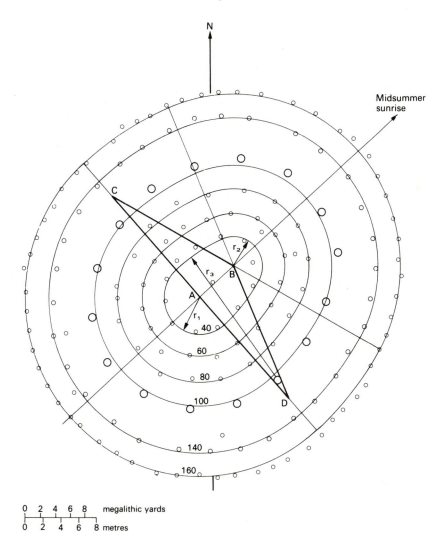

Fig. 10. Plan of Woodhenge. The numbers beside the rings are the perimeters in Megalithic Yards. From J. E. Wood: Sun, Moon and Standing Stones, p. 45

case of the relatively small distance AB between well-defined points A and B we may be pretty sure that the distance is just $6\ MY$, but are AC and CB exactly $17\frac{1}{2}$ and $18\frac{1}{2}\ MY$?

These doubts are justified as long as large distances are concerned, especially when these distances are not integer but only half-integer multiples of the Megalithic Yard. But if one considers the totality of the evidence from large and small distances measured very carefully in Brittany (Bretagne), in Southern England and in Scotland, which are all integer or half-integer multiples of one and the same fundamental length, one gets

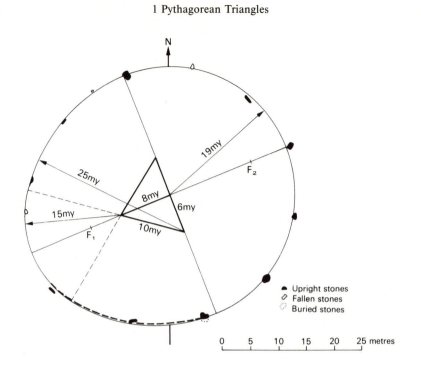

Fig. 11. Geometry of the ring at Clava. The plan is based on one by A. Thom, and the dimensions on the figure are his interpretations of the distances in Megalithic Yards. From J. E. Wood: Sun, Moon and Standing Stones, p. 50

the conviction that Thom's Megalithic Yard is well established. Of course, the last decimal in the figure 0.829 is uncertain: it was kept in order to avoid rounding errors, but the existence of a common standard measure of nearly 83 cm used in these three countries cannot well be doubted.

In my opinion, it is very difficult to find Pythagorean triples like (3,4,5) or (5,12,13) or (12,35,37) empirically. Among the numerous right-angled triangles having integer perpendicular sides there are very few having an integer hypotenuse. As we shall see, there are good reasons to assume that these triples have been calculated by the arithmetical methods explained at the beginning of this chapter.

Megalithic Architecture in Egypt

Megalithic architecture requires a considerable amount of technological skill. Large stones must be quarried, transported over considerable distances, and dressed to fit together. Lintels must be lifted to form huge trilithons.

In the fourth millennium B.C., we find megalithic constructions not only in Portugal, Spain, and Brittany, but also on the island of Malta. In the third millennium huge passage graves were built in Egypt: the pyramids. In the second millennium, passage graves with large trilithons were built in Mycenae. To erect these trilithons and to construct the large Egyptian temples that were built in the second millennium B.C., the same technological skill is required which we admire in Stonehenge II.

I have already explained why I don't believe in independent inventions, apart from the very few cases in which we happen to know that a scientific or technological invention was made twice. In accordance with this heuristic principle, I now propose the following working hypothesis:

The technical skill necessary for megalithic architecture was developed only once, and the fashion of megalithic architecture spread from just one centre.

Note that Greek temple architecture, Romanesque church building, Gothic architecture, and Renaissance architecture all spread from one centre. So my working hypothesis is certainly not unrealistic.

This hypothesis implies that the English henges and the Egyptian pyramids and temples were derived from one common origin. There are several additional arguments in favour of this conclusion, namely:

1. The passage graves in Western Europe were burying places of a ruling class, and so were the Egyptian pyramids.

2. In England as well as in Egypt the ruling class was able to force or persuade the people to perform the enormous amount of work necessary for the erection of megalithic buildings.

3. In England the henges were religious meeting places, and so were the temples in Malta and Egypt.

4. Some megalithic monuments in England and Ireland were oriented towards the midwinter or midsummer sunrise, and the same thing holds for Egyptian temples. For instance, the great temple of Amon-Ra at Karnak was directed towards the midwinter sunrise. See G.S. Hawkins: Beyond Stonehenge. New York 1973, Chapter 11.

5. In Stonehenge we find an arrangement of "causeway postholes" which can only be explained as a result of systematic observation of the moon's rising during several years, probably for the purpose of predicting eclipses (see J.E. Wood: Sun, Moon, and Standing Stones, p. 101). Hence we may conclude that the megalith builders practiced some kind of "horizon astronomy" based on observations at or near the horizon. Now the Egyptians also observed stars near the horizon. About 2000 B.C., under the Middle Kingdom, they developed a rather primitive theory of the rising and setting of fixed stars. I have explained this theory in Chapter 1 of my book "Science Awakening II" (Leyden 1974).

6. The circles and ellipses in the English henges were probably constructed by means of stretched ropes. In Egypt too, there was a class of "rope-stretchers", called in Greek "Harpedonaptai", who played an important rôle in the initial ceremonies connected with the foundation of temples. The Greek and Egyptian testimonies concerning the Harpedonaptai were thoroughly discussed by S. Gandz in his paper "Die Harpedonapten oder Seilspanner" (Quellen und Studien Gesch. der Math. B 1, 1930,

p. 255–277). In an inscription describing the foundation of a temple at Abydos by Sethos I (1300 B.C.), the goddess is made to speak to the king thus: "You were with me in your function as Rope-Stretcher." Still earlier, Thutmose III (1500 B.C.) is said to have spanned the rope towards the sun-god Amon at the horizon.

In his paper just quoted, Gandz has discussed a fragment of Democritos according to which the harpedonaptai were experts in "composing lines". Now this is just what the builders of stone rings in England and Scotland did: they composed arcs of circles to form large rings, and they constructed ellipses.

These arguments are probability arguments. No single one of them is conclusive, but combined they provide a strong support of my general thesis.

Finally, I may note that the Egyptians too had some knowledge of Pythagorean triples. In the Middle Kingdom papyrus Berlin 6619, published by Schack-Schackenburg in 1920 (Zeitschrift für ägyptische Sprache 38, p. 138 and 40, p. 65) the first problem reads:

> A square and a second square, whose side is one-half and one-quarter of that of the first square, have together an area of 100. Show me how to calculate this.

If the two unknown sides are called x and y, the problem is equivalent to the pair of equations

$$y = \tfrac{3}{4} x$$

$$x^2 + y^2 = 100.$$

The solution is $x=8$, $y=6$. The sum of the squares is again a square, namely 100, and this fact is used in the solution, for the text says "take the square root of the given number 100; it is 10". So we have

$$x^2 + y^2 = z^2$$

with $x=8$, $y=6$, $z=10$. The triple (8,6,10) is twice the well-known triple (4,3,5).

The occurrence of this triple cannot be used as an additional argument in favour of my thesis, for it is quite possible that the Egyptians learnt about Pythagorean triples from the Babylonians. A triangle is not mentioned in the text.

The Ritual Use of Pythagorean Triangles in India

In altar constructions described in the Āpastamba Śulvasūtra (see Seidenberg, Archive for History of Exact Sciences 18, 1978, p. 323) several Pythagorean triangles are used, namely (see Fig. 12):

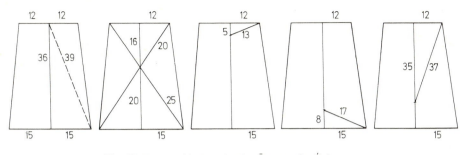

Fig. 12. Trapezoid altars in the Āpastamba Śulvasūtra

(3, 4, 5) and its multiples (12,16,20) and (15,20,25)
(5,12,13) and its multiple (15,36,39)
(8,15,17) and (12,35,37).

These triangles are combined to form larger figures in the same way as in the "Metrica" of Heron of Alexandria. We have here one more argument in favour of the hypothesis of a common origin of the use of Pythagorean triangles in Greece and India.

Leaving aside multiples, we see that the triples (3,4,5), (5,12,13), (8,15,17), (12,35,37) were used in India for ritual purposes. Now it is very remarkable that three of these triples, namely

$$(3,4,5), \quad (5,12,13), \quad (12,35,37)$$

were also used in England and Scotland for ritual purposes, namely for the construction of religious meeting places. Moreover, the triples

$$(3,4,5) \quad \text{and} \quad (5,12,13)$$

are just the first two that can be obtained from the rule of Pythagoras, whereas

$$(4,3,5), \quad (8,15,17), \quad (12,35,37)$$

are the lowest primitive triples that can be obtained from the rule of Plato.

I am convinced that these coincidences are not accidental. In the Neolithic Age there must have been a doctrine of Pythagorean triples and their ritual applications.

Part C

On Proofs, and on the Origin of Mathematics

Geometrical Proofs

Proofs of geometrical propositions are found in the earliest known mathematical texts in India and China as well as in Greece. I shall give some examples.

In the Āpastamba Śulvasūtra, an altar is described in the form of an isosceles trapezium with its eastern base 24 units, its western 30, its width 36. The text says that the area is 972 square units. This is proved as follows (see Fig. 13).

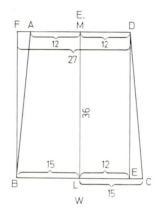

Fig. 13. Area of a trapezium

One draws (a line) from the southern *amsa* (*D* in Fig.) toward the southern *srōni* (*C*), (namely) to (the point *E* which is) 12 (*padas* from the point *L* of the *prsthya*). Thereupon one turns the piece cut off (i. e. the triangle *DEC*) around and carries it to the other side (i. e. to the north). Thus the *vedi* obtains the form of a rectangle. In this form (*FBED*) one computes its area.

The translation is taken from Seidenberg's paper "The Origin of Mathematics", Archive for History of Exact Sciences 18, 1978, p. 332. I fully agree with Seidenberg's comment: "The striking thing is that here we have a proof."

The earliest Chinese text on astronomy and mathematics is the *Chou Pei Suan Ching*. In Needham's book "Science and Civilization in China", Vol. 3, the Chinese title is translated as "Arithmetical Classic of the Gnomon and the Circular Paths of Heaven". The text was written in the Han-period, but it was certainly based on earlier traditions.

In this classical text, a proof of the "Theorem of Pythagoras" is presented. The proof is worked out only for the (3,4,5) triangle, but the idea of the proof is perfectly general. In Needham's rather free translation of the difficult text the proof goes as follows:

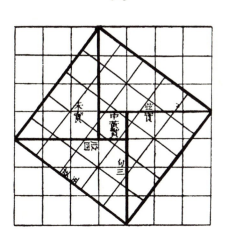

Fig. 14. The proof of the Pythagoras Theorem in the Chou Pei Suan Ching. From J. Needham: Science and Civilisation in China, Vol. 3

Thus, let us cut a rectangle (diagonally), and make the width 3 (units) wide, and the length 4 (units) long. The diagonal between the (two) corners will then be 5 (units) long. Now after drawing a square on this diagonal, circumscribe it by half-rectangles like that, which has been left outside, so as to form a (square) plate. Thus the (four) outer half-rectangles of width 3, length 4, and diagonal 5, together make two rectangles (of area 24); then (when this is subtracted from the square plate of area 49) the remainder is of area 25. This (process) is called "piling up the rectangles".

The calculation starts with a large square, having a side $3+4=7$ and an area $7^2=49$. Now, four triangles, making together two rectangles of area $3 \times 4 = 12$ are subtracted, and what remains is a square of side 5:

$$49 - 24 = 25 = 5^2.$$

In the general case, when the width and the length are called a and b and the diagonal c, the same method of proof would yield

$$c^2 = (a+b)^2 - 2ab = a^2 + b^2.$$

The mathematician who invented this proof would have to know the identity

$$(a+b)^2 = a^2 + b^2 + 2ab.$$

The geometrical proof of this identity is very easy: we find it in Euclid's Elements (Proposition II,4) as well as in the Algebra of Al-Khwārizmī (see

Fig. 15). The Babylonians too knew this identity (see my Science Awakening I, p. 68–69). The Chinese might have read off the same identity from their own drawing (Fig. 14): two squares of areas $a^2 = 4^2$ and $b^2 = 3^2$ at the bottom together with two rectangles of area ab each make up the large square of area $(a+b)^2$.

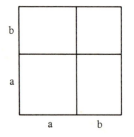

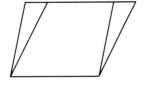

Fig. 15. Geometrical proof of the formula for $(a+b)^2$

Fig. 16. Euclid's proof of the equality of parallelograms

Note that the Chinese proof of the Theorem of Pythagoras is quite similar to the proof previously quoted from the Āpastamba Śulvasūtra, in which a triangle is cut off from the trapezium, and an equal triangle is added on the other side, and also to several proofs in Euclid's Elements. For instance, when Euclid wants to prove that two parallelograms on the same base having equal heights are equal, he adds a triangle to one of the parallelograms, and subtracts an equal triangle on the other side (see Fig. 16).

The Chinese proof may be reformulated in the same way. One starts with the sum of the two squares a^2 and b^2, one adds two rectangles of area ab each, thus obtaining the square $(a+b)^2$, and one subtracts four triangles of area $\frac{1}{2}ab$ each. What remains is the square c^2.

I suppose that proofs of this kind formed a part of an oral tradition current in the Neolithic Age, and that the proofs in the Sanskrit, Greek, and Chinese written texts were ultimately derived from this oral tradition.

Looking again at the drawing accompanying the Chinese proof, we may notice that four triangles and a central square are bounded by heavy lines. This square and its surrounding triangles are not used in the text. It seems to me that the purpose of the heavy lines can only be to suggest another proof of the theorem. The area of the central square is $(a-b)^2$, and the areas of the four surrounding triangles are together $2ab$. So we have

$$c^2 = (a-b)^2 + 2ab = a^2 + b^2.$$

The first proof was based on the identity

(15) $$(a+b)^2 = a^2 + b^2 + 2ab$$

and the second on the identity

(16) $$(a-b)^2+2ab=a^2+b^2.$$

The identity (15) is Euclid's Proposition II,4, and (16) is Euclid's II,7.

I am convinced that the excellent neolithic mathematician who discovered the Theorem of Pythagoras had a proof of the theorem, but I don't know whether he knew one of the Chinese proofs or both, for there are still more possibilities, as we shall see presently.

Euclid's Proof

In Euclid's Elements, at the end of the first book, a proof of the theorem is given which is based on a division of the square c into two rectangles $p \cdot c$ and $q \cdot c$. Euclid proves that $p \cdot c$ is equal to a^2, and $q \cdot c$ equal to b^2. Hence the theorem follows.

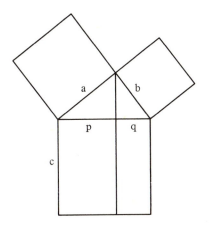

Fig. 17. Euclid's proof of the theorem of Pythagoras

Euclid's proof of the equalities $pc=a^2$ and $qc=b^2$ is rather complicated, because it avoids proportions and uses only transformations of areas of triangles. If one uses proportions, the proof of the two equalities is easy. One has, obviously,

$$p:a=a:c, \quad \text{hence } pc=a^2,$$
$$q:b=b:c, \quad \text{hence } qc=b^2.$$

I guess Euclid (or one of his predecessors) knew this simple proof from an earlier tradition and replaced it by a more complicated proof because he wanted to avoid, at this early stage of his exposition, the use of proportions.

Naber's Proof

Dr. Naber, a teacher of mathematics in the little town Hoorn in Holland, invented a very nice proof of the theorem[4]. His starting point is the simple observation that the line CD, drawn from the top of the triangle perpendicular to the base, divides the trangle ABC into two *similar* triangles CBD and ACD. This fact was known to Euclid, for he proves it as Proposition VI,8:

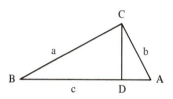

Fig. 18. Euclid's proposition VI, 8

If in a right-angled triangle a perpendicular be drawn from the right angle to the base, the triangles adjoining the perpendicular are similar both to the whole and to one another.

Now the areas of similar figures, similarly described on their bases, are in the same proportion as the squares on their bases (see Fig. 19).

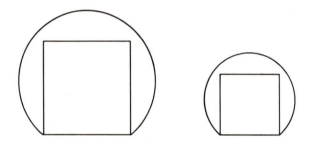

Fig. 19. Proportionality of circle segments

This fact was known long before Euclid, for Hippocrates of Chios stated it for similar circle segments, and he based his quadrature of lunulae on this statement (see e.g. my Science Awakening I, p. 131–135). So, since the triangle described on the base c is the sum of the similarly described triangles on the bases a and b, the square on c is equal to the sum of the squares on a and on b. More generally, the area of any plane figure described on c is equal to the sum of the areas of similar figures described on a and b.

This generalization of the Theorem of Pythagoras was known to Euclid, for his Proposition VI,31 reads:

---

4 H. A. Naber: Das Theorem des Pythagoras, Haarlem 1908.

In right-angled triangles any figure on the side subtending the right angle is equal to the
similar and similarly described figures on the sides containing the right angle.

So Euclid had in his hands all elements of Naber's proof. It is easy to
see why he did not give the full proof. It was not necessary, because he had
proved the theorem already in I,14.

By Naber's proof, the Theorem of Pythagoras becomes immediately
evident. One only has to look at the three triangles and to see that they are
similar, and the theorem follows at once. As Naber noted, the Greek word
theorema is derived from *theorein:* to look at.

We have now, all in all, four proofs, namely two Chinese proofs and
two proofs indicated between the lines of Euclid's text. Any one of these
proofs could well have led to the discovery of the theorem in the Neolithic
Age.

Astronomical Applications of the Theorem of Pythagoras?

In the megalithic stone rings we have found clear indications of astronomical as well as mathematical activity. In all ancient civilizations, astronomy and mathematics always go together. Oinopides of Chios, one of the
earliest Greek astronomers, invented a geometrical construction "because
he thought it useful in astronomy" (Proclos, Commentary to the first book
of Euclid's Elements, p. 220 in the translation of Morrow). Later mathematicians such as Eudoxos, Archimedes, and Apollonios were at the same
time astronomers. In China, a proof of the Theorem of Pythagoras is contained in an astronomical treatise, the "Arithmetical Classic of the Gnomon and the Circular Paths of Heaven".

This coincidence of mathematical and astronomical activity is not accidental. Even the most primitive astronomical theories are based on geometrical ideas and require arithmetical computations. For examples see my
"Science Awakening II".

In neolithic mathematics, the Theorem of Pythagoras plays a central
rôle, so it is only natural to ask: Is it possible that this theorem had a useful
application in neolithic astronomy? If it had, we could explain why a proof
of the theorem was included in a Chinese astronomical treatise.

In Ptolemy's Almagest, the Theorem of Pythagoras plays an important
rôle in the foundations of trigonometry and the computation of sines, and
it is also used for calculating the duration of an eclipse. Most probably,
trigonometry was not invented before Apollonios (circa 200 B.C.), but the
application to eclipses is independent of trigonometry. How does it
work?

During a lunar eclipse, the sun remains nearly stationary at a certain
point on the ecliptic, and the moon moves in an orbit which may be above
or below the ecliptic at a certain distance. Suppose, one knows this distance *d,* and one also knows that the eclipse begins and ends when the
moon's disc just touches a particular circular shadow, the center of the
shadow being a point *A* on the ecliptic just opposite the sun (see Fig. 20):

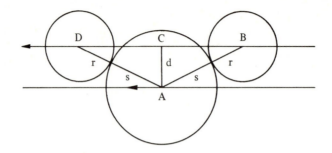

Fig. 20. The moon's path during a lunar eclipse

It is not necessary to know that the circular shadow is the shadow of the spherical earth.

If the moon's apparent diameter is r, and the radius of the shadow s, the distance between the centres of the moon and the shadow at the beginning and end of the eclipse is $r+s$. If d and $r+s$ are known, the third side BC of the right-angled triangle ABC can be calculated by the Theorem of Pythagoras, and hence $BD = 2BC$ is known. The velocity of the moon relative to the sun is also known: it is one full revolution in $29\frac{1}{2}$ days. So, the time the moon needs to cover the distance BD can be calculated. This is the method used by Ptolemy and by all Hindu and Islamic astronomers.

If we assume that the astronomers of the Neolithic Age already used this method, we can explain why the Theorem of Pythagoras was taught and transmitted to the Babylonians, the Greeks, the Hindus, and the Chinese. Of course, this is only a hypothesis, but as far as I can see it is the only hypothesis that explains the observed facts.

Why Pythagorean Triangles?

The knowledge of the Theorem of Pythagoras does not necessarily imply the calculation of Pythagorean triangles. For the construction of right angles no Pythagorean triangles are needed. If one makes $AB = BC$ and $AD = CD$, the angle at B will be a right angle (Fig. 21). This construction, which was used by Euclid, can easily be performed by means of stretched ropes.

Euclid continually uses the Theorem of Pythagoras, but he does not say a word about Pythagorean triples or triangles. Our engineers and physicists learn the theorem at school and sometimes use it, but they have no use for Pythagorean triples.

Now the problem arises: How can we explain the fact that methods for finding Pythagorean triples found their way into Babylonian, Hindu, Greek, and Chinese texts? For what purpose were these methods invented and transmitted to later generations?

The clearest exposition of these methods is found in the Chinese "Nine Chapters". So it is only natural to ask: For what purpose were these triples used in the "Nine Chapters"?

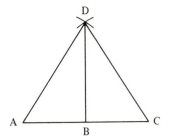

Fig. 21. Construction of a right angle

The final chapter of this treatise consists of a sequence of problems with solutions, all concerning right-angled triangles. The problems will be discussed in the next chapter. In each of these problems, two numbers are given, for instance one side and the sum of the other two. Sometimes the solution can be found by rational operations (addition, subtraction, multiplication, and division), but sometimes the extraction of a square root is necessary. Now all given numbers are chosen in such a way that the square root is a rational number. In order to attain this, the author of the problems had to take care that his triangles had rational sides, that is, he had to know how to find Pythagorean triangles.

As we shall see in the next chapter, the Chinese collection of problems is closely related to Babylonian collections, in which problems of exactly the same kind were posed and solved by the same methods. The similarity is so close that a pre-Babylonian common origin must be assumed.

I now suppose that Pythagorean triples were invented just for this purpose: to compose problems with rational solutions. There must have been a tradition of teaching mathematics by means of well-chosen sequences of problems with solutions, a tradition which originated somewhere in neolithic Europe and which spread towards Babylonia, Greece, and China.

The Origin of Mathematics

We have seen that there are so many similarities between the mathematical and religious ideas current in England in the Neolithic Age, in Greece, in India, and in China in the Han-period, that we are bound to postulate the existence of a common mathematical doctrine from which these ideas were derived.

Can we make a reasonable conjecture about the place of origin of this mathematical doctrine?

In the English henges we find Beaker Pottery, or more precisely "Glockenbecher", as the German scholars call this kind of pottery (see Fig. 22):

GBK

Niederlande

Fig. 22. Beaker pottery, Netherlands

The domain in which Beaker Pottery is found is shown in Fig. 23:

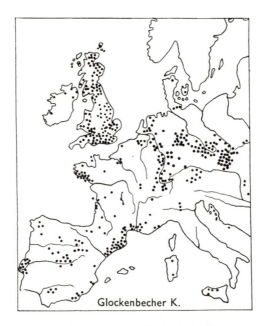

Glockenbecher K.

Fig. 23. "Glockenbecherkultur"

Figures 22 and 23 are taken from the work of Jan Filip: Enzyklopädisches Handbuch zur Ur- und Frühgeschichte Europas I (1966). It is seen that the domain extends from Portugal, Brittany, and the British Isles to Czechoslovakia and even further east.

The Beaker people, who built Stonehenge II, Woodhenge and other henges, came to England from the continent shortly after 2500 B.C. It is possible that they already spoke an Indo-European language. In any case, they lived in a region where Indo-European languages were spoken at an early date.

The Indo-European languages are connected with a perfect decimal counting system, including a method of designating fractions. The English expression "the fifth parth" corresponds to the Greek τὸ πέμπτον μέροσ. This number system is an important cultural achievement and an excellent basis for teaching arithmetic and algebra. Also, the religions of ancient Indo-European populations have so much in common that the existence of an Indo-European religion can hardly be doubted. Hence, if we find quite similar ideas about the ritual importance of geometrical constructions in Greece and India, and the same set of Pythagorean triangles with ritual applications in England and India, and the same geometrical constructions in Greece and India, the conclusion that these religious and mathematical ideas have a common Indo-European origin is highly probable.

Chapter 2

Chinese and Babylonian Mathematics

Part A

Chinese Mathematics

The Chinese "Nine Chapters"

The Chinese collection "Nine Chapters on the Mathematical Art" *(Chiu Chang Suan Shu)* was composed during the Han-period, that is, between 206 B.C. and A.D. 221. According to Liu Hui (third century A.D.), who wrote a commentary to the Nine Chapters, the work was based on an earlier collection, which was burnt in the time of the emperor Ch'in Shih Huang (221–206 B.C.). Remnants of the earlier collection were later recovered and arranged in nine chapters, says Liu Hui. See the article LIU HUI by Ho Peng-Yoke in the Dictionary of Scientific Biography.

The Nine Chapters are available in a Russian translation by R.I. Berezkina (Istorika-matematicheskie issledovaniya 10, 1957) and in a German translation by K. Vogel (Vieweg, Braunschweig 1968). By the kind permission of Donald Blackmore Wagner, I shall here use his English translation.

The contents of the Nine Chapters may be summarized thus:

Chapter 1
Operations on Fractions *m/n*
Measurements of Areas
Chapter 2
Exchange of Goods and Money
Indeterminate Equations
Chapter 3
Partitions in a Given Proportion
Commercial Problems

I shall now discuss some of the most interesting parts of the Nine Chapters.

The Euclidean Algorithm

The rules for addition, subtraction, multiplication, and division of fractions m/n, explained in Chapter 1, are easy enough: they agree with the rules we learn at school. More interesting is the rule for simplifying fractions.

Let a mixed fraction m/n be given. If m and n can be halved, they are halved, says the text. If not, "lay out the two numbers on the counting board". Next "diminish the numbers by alternate subtraction until you get equal numbers". Finally, divide by these equal numbers.

The text gives the example 49/91. We are instructed to lay out

$$\begin{array}{cccccccc} 49 & 49 & 7 & 7 & 7 & 7 & 7 & 7 \\ 91 & 42 & 42 & 35 & 28 & 21 & 14 & 7 \end{array}$$

and to divide numerator and denominator by 7.

Is this not wonderful? We have here a completely clear description of the "Euclidean Algorithm".

Note that for performing calculations with fractions the Euclidean Algorithm is not strictly necessary. The text explains the addition and multiplication of fractions by means of rules like

$$\frac{a}{b} + \frac{c}{d} + \frac{e}{f} = \frac{adf + cbf + ebd}{bdf}$$

$$\frac{a}{b} \cdot \frac{c}{d} \cdot \frac{e}{f} = \frac{ace}{bdf}$$

and one can perfectly well leave the resulting fractions unsimplified, or else simplify them only if one finds obvious common divisiors. In our case 49/91 it is easy to see that 49 is 7 times 7, and that 91 is divisible by 7. So the mention of the algorithm is not a logical or didactical necessity: it is just an addition by a systematically-minded teacher, who wanted to teach a never-failing method.

Let us now compare the Chinese method of simplifying fractions m/n with the Greek method of reducing a ratio $m:n$ to lowest terms, which is taught in Book 7 of Euclid's Elements. The methods are essentially the same. The Greeks as well as the Chinese reduce the pair of numbers (m,n) by alternate subtractions until they become equal, and then they divide m and n by the resulting common divisor.

At the time of Plato, the Greek mathematicians did not recognize fractions: they regarded the One as indivisible (Plato, Republic 525 E). Instead of fractions m/n they used ratios $m:n,$ but of course they knew very well how to handle fractions. This is seen from the work of Archimedes, Eratosthenes, and Diophantos. Archimedes uses the numbers 3 1/7 and 3 10/71 in his "Measurement of the Circle". Eratosthenes uses the fraction 11/83 in his estimation of the obliquity of the sun's orbit. Diophantos allows fractions as solutions of his indeterminate equations.

The Greeks had a complete decimal counting system, and they performed calculations with fractions in the same way as the Chinese did. The Greeks and Chinese both used the counting board, and they ordered large numbers according to powers of 10 000.

It is extremely improbable that all this, including the Euclidean Algorithm, was invented independently by the Greeks and the Chinese. Remember that they both had methods to calculate Pythagorean triples, and that the Chinese method is essentially the same as the Greek method. Pythagorean triangles did not form an essential part of either Greek or Chinese mathematics: they were just playful additions, which enabled Diophantos and the author of the Nine Chapters to propose nice little problems about right-angled triangles.

It is also very improbable that all this came from the Greeks to the Chinese. The rules for adding and multiplying fractions and for simplifying them by means of the Euclidean Algorithm are not found in any extant Greek treatise, and the general rule for computing Pythagorean triangles is found only in the very late treatise of Diophantos. So the only reasonable explanation of the many similarities seems to be, once more, the hypothesis of a common origin.

Areas of Plane Figures

In Chapter 1 of the Nine Chapters we find rules for the calculation of areas of rectangles, triangles, trapezia, circles, circle segments, and circle sectors. The rules for rectangles, triangles, and trapezia are all correct. For the circle sector (Problem 33) the rule says: Multiply the diameter by the arc, and divide by 4. This rule is also correct. It shows that the author of the problem was completely aware of the relation between the arc, the radius and the area of a circle or circle sector:

(1) $\qquad\qquad$ Area $= \frac{1}{2}$ Radius $\times$ Arc.

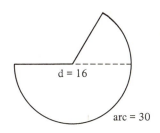

Fig. 24. Chinese example of a circle sector

The first known proof of this relation was given by Archimedes in his "Measurement of the Circle".

If the relation (1) is applied to a full circle, one obtains

(2) $\qquad\qquad$ Area $= \frac{1}{2}$ Radius $\times$ Circumference

or, equivalently,

(3) $\qquad\qquad$ Area $= \frac{1}{4}$ Diameter $\times$ Circumference.

The text presents both rules (2) and (3), and also a third rule:

(4) $\qquad\qquad$ Area $= \frac{3}{4}$ Diameter $\times$ Diameter

and a fourth

(5) $\qquad\qquad$ Area $= \frac{1}{12}$ Circumference $\times$ Circumference.

If we compare (4) and (5) with (3), we see that the circumference of a circle was equated to 3 times the diameter. The same ratio was adopted in Babylonian texts, and the rule (1) was also known to the Babylonians (see A. Seidenberg: On the Area of a Semi-Circle, Archive for History of Exact Sciences 9, p. 171–211, especially footnote 26).

Most remarkable is the Chinese rule for computing the area of a segment of a circle (Chapter 1, Problems 35 and 36). The rule reads:

Multiple the chord by the arrow. Also multiply the arrow by itself. Add and divide by two.

If the chord is s and the arrow h (see Fig. 25), the Chinese rule can be written as

(6) $\qquad\qquad\qquad A = \frac{1}{2}(sh + h^2).$

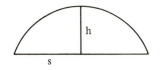

Fig. 25. Chord and arrow of a circle
segment

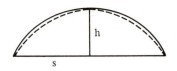

Fig. 26. Parabola inscribed in a circle
segment

If the segment is a semi-circle, the formula (6) is equivalent to (4), which is approximately correct. On the other hand, if h is small as compared with s, the circle segment may be replaced, in a very good approximation, by a segment of a parabola (see Fig. 26). The area of this segment is according to Archimedes

$$(7) \qquad\qquad A' = \tfrac{2}{3} s h.$$

For small h, the quotient A/A' tends to $3/4$, so the Chinese approximation (6) is not good.

It is very curious that the same inaccurate formula (6) also occurs in the "Metrica" of Heron of Alexandria (p. 72–75 in the edition of Schöne) and in a papyrus from Cairo written in the third century B.C. and published by R.A. Parker[5]. Heron writes the formula (6) in the equivalent form

$$(8) \qquad\qquad A = \tfrac{1}{2}(s + h)h$$

and ascribes it to "the ancients". Heron's text and the Cairo papyrus will be discussed more amply in Chapter 6.

As a general principle, if one finds one and the same *correct* rule of computation in several civilizations, one always has to take into account the possibility of independent invention, but if the rule is *incorrect,* independent invention is next to impossible. Therefore we are bound to suppose that the Chinese formula (6) and the equivalent formula (8), used in the Cairo papyrus and mentioned by Heron, were derived from a common origin.

According to Parker, the Demotic papyrus shows clear traces of Babylonian influence. The problems and solutions in the papyrus are of just the same kind as those we find in Babylonian problem texts. However, the false rule (8) is not found in any extant Babylonian text, as far as I know.

Volumes of Solids

In Chapter 5 of the Nine Chapters, the first seven problems are practical: they are concerned with volumes of walls and dams and with the num-

5 R.A. Parker: Demotic Mathematical Papyri, Brown University Press, Providence R.I., and Lund Humphreys, London 1972.

bers of workmen needed for construction purposes. Next follows a sequence of purely theoretical problems on volumes of solids. Here I shall
restrict myself to the following solids:

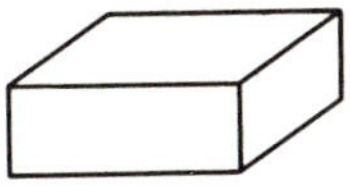

Fig. 27

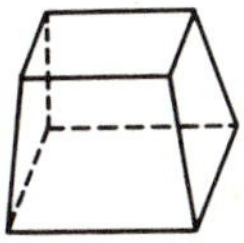

Fig. 28

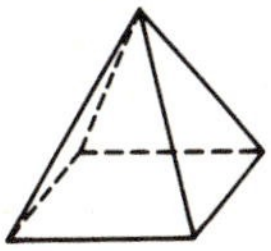

Fig. 29

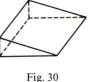

Fig. 30

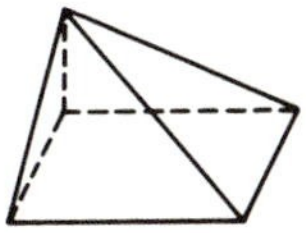

Fig. 31

> *Fang pao tao:* rectangular block (Fig. 27),
> *Fang t'ing:* truncated pyramid with square base (Fig. 28),
> *Fang chui:* pyramid with square base (Fig. 29),
> *Ch'ien tu:* wedge (Fig. 30),
> *Yang ma:* pyramid with square base, in which one of the edges
> is perpendicular to the base plane (Fig. 31).

In all these cases, the text gives correct prescriptions for calculating the
volumes. In particular, the volume of the truncated pyramid is given as

(9) $V = \frac{1}{3}(a^2 + ab + b^2)h,$

where a is the side of the top square, b the side of the base square, and h
the height. See Donald B. Wagner: A Chinese Derivation of the Volume of
a Pyramid. Historia Mathematica 6, p. 164–188 (1979).

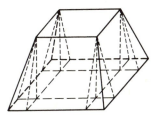

Fig. 32. Division of a truncated pyramid into pieces

The last two solids, *ch'ien tu* and *yang ma,* are of little practical importance in themselves, but if one divides the truncated pyramid by vertical
dividing planes as indicated in Fig. 32, the parts are:

one central block *(fang pao tao)*,
four wedges *(ch'ien tu)*,
four pyramids *(yang ma)*.

Liu Hui, the third century commentator of the Nine Chapters, describes just this division and uses it in his proof of (9). He works out the proof in the special case $h=1$, $a=1$, $b=3$ only, but the idea of the proof is perfectly general. To understand the proof, one must know that his unit of length is the *ch'ih,* a "foot" of about 23 cm, and that his units of area and of volume, the square foot and the cube foot, are also denoted by the same word *ch'ih.*

The proof begins thus:

The product of (the sides of) the upper and lower squares is 3 *ch'ih,* and multiplying this by the height gives 3 *ch'ih.* This means there is obtained one central cube and one each of the wedges at the four sides.

The meaning is: the central block and the four wedges can be united to form one large block with sides $a=1$, $b=3$, $h=1$ and volume $abh=3$. Hence we have

$$(10) \qquad abh = \text{central block} + 4 \text{ wedges}.$$

In the same way Liu uses the fact that the central block and twice the four wedges and three times the four pyramids can be united to form one large block with volume b^2h:

$$(11) \qquad b^2h = \text{central block} + 2 \text{ (sum of the 4 wedges)}$$
$$+ 3 \text{ (sum of the 4 pyramids)}.$$

Liu's third equation is

$$(12) \qquad a^2h = \text{central block}.$$

Adding the three equations, Liu concludes

$$(13) \qquad abh + b^2h + a^2h = \quad 3 \text{ (central block)}$$
$$+ 3 \text{ (sum of the 4 wedges)}$$
$$+ 3 \text{ (sum of the 4 pyramids)}.$$

Dividing by 3, he obtains the desired result (9).

In this proof, it is presupposed that the volume of a pyramid with square base a^2 and height h is $a^2h/3$. In another section of his commentary, Liu gives an ingenious proof of this rule. His proof will be reproduced in our Chapter 6.

It is instructive to compare Chapter 5 of the "Nine Chapters" with the Old-Babylonian text BM 85 194 (Neugebauer, Math. Keilschrifttexte I, p. 142–193). Just like Chapter 5, this text begins with problems about volumes of dams and walls and the number of workmen needed to build them.

In most cases, in the Chinese as well as in the Babylonian text, the cross-section of the wall or dam is supposed to be a trapezium, and the volume is calculated as the product of the length and the area of the cross-section.

The problem treated in § 9 of Neugebauer's commentary is the calculation of the volume of a truncated cone. The height is $h=6$, the lower circumference $u=4$, the upper circumference $v=2$. The areas of the two circles are calculated as

$$A=u^2/12 \quad \text{and} \quad B=v^2/12.$$

Next, the volume V is calculated as

(14) $$V=\tfrac{1}{2}(A+B)\cdot h.$$

The rule $A=u^2/12$ is also found in the Nine Chapters, as we have seen. The formula (14) is incorrect. The correct formula (correct, that is, if one assumes $\pi=3$) is given in Chapter 5 of the Nine Chapters (Problem 11):

(15) $$V=[(uv+u^2+v^2)/36]\cdot h.$$

In the case of the truncated pyramid, the situation is more complicated. In the text BM 85210 (Neugebauer, MKT I, p. 219–233), which is closely related to our text BM 85194, the volume of a truncated pyramid with square base a^2, top surface b^2 and height h, is calculated by the incorrect formula

(16) $$V=\tfrac{1}{2}(a^2+b^2)\cdot h$$

corresponding to (14). This seems to be the Babylonian standard method for computing the volumes of truncated cones and pyramids. A similar method was applied to a dam having a variable cross-section in BM 85194 (see Neugebauer, MKT I, p. 165). However, in the same text, the volume of a truncated pyramid is calculated by means of a different prescription, which may be written as

(17) $$V=[([a+b]/2)^2+\ldots]\cdot h$$

where … represents a doubtful term, the text being damaged. Neugebauer proposed to read this term as

$$\tfrac{1}{3}([a-b]/2)^2.$$

If this interpretation is accepted, the formula (17) would be equivalent to the correct Chinese formula (9). On the other hand, the doubtful term in (17) can also be interpreted as

$$([a-b]/2)^2,$$

and then the formula (17) would be equivalent to the incorrect formula (16). See my Science Awakening I, p. 76.

Conclusion: It is doubtful whether the Babylonians knew the correct Chinese formula (9), but in any case they considered problems of just the same kind as the Chinese, namely: how to calculate volumes of truncated cones and pyramids and how to calculate the number of workmen needed for earthwork. The Babylonians and the Chinese also used the same formula

$$(18) \qquad\qquad A = \tfrac{1}{12} u^2$$

to calculate the area A of a circle with given circumference u. It is clear that the Babylonian and Chinese problem texts are closely related.

The Moscow Papyrus

The correct formula for the volume of a truncated pyramid was known not only in China, but also in ancient Egypt. Under the Middle Kingdom (circa 2050–1800 B.C.) the "Moscow Papyrus" was written. It was published by W. W. Struve: Mathematischer Papyrus des Staatlichen Museums der schönen Künste in Moskau, Quellen und Studien Geschichte der Mathematik, Vol. A1 (1930). In this papyrus, the volume of a truncated pyramid is calculated by just the same rule as in the Nine Chapters, namely by the correct formula (9).

The Moscow Papyrus is a problem text of the same kind as the Chinese and Babylonian problem texts. The fact that one and the same rule for calculating the volume of a truncated pyramid occurs in the Egyptian papyrus as well as in the Nine Chapters is a strong argument in favour of a common origin.

The question now arises: Is it possible that this origin was a Babylonian text? It is possible, but Babylonian mathematics and astronomy are well known to us from hundreds of cuneiform texts. In all these texts, the correct rules for computing the volumes of truncated cones and pyramids are not found, but the rule for pyramids was known to the Egyptians and Chinese, and the rule for cones was known to the Chinese. So it seems that we have to assume a *pre-Babylonian* common source.

Similarities Between Ancient Civilizations

In comparing the civilizations and in particular the mathematical abilities of the megalith builders in Western Europe, the Greeks, the Egyptians, the Babylonians, the Hindus, and the Chinese, we have found many similarities. In the following summary, I shall restrict myself to the most important facts.

1°. In all countries just mentioned we find:
Agriculture yielding a surplus (the conditio sine qua non for any higher civilization), and a highly civilized Ruling Class living on this surplus.

2°. In the West, in Greece, Phoenicia, and Egypt we find:
Shipbuilding and Open Sea Navigation.

3°. In the West, in Greece, Asia Minor, Iran, India, and China:
Indo-European Languages (Hittite in Asia Minor, Tocharian in China).

4°. In Greece, Egypt, India, and China:
A completely developed decimal counting system, and rules for operations with fractions m/n (in Egypt only 2/3, 3/4, and $1/n$).

5°. In Greece and China:
The Euclidean Algorithm and the Counting Board.

6°. In the West, in Egypt and Babylon:
Horizon Astronomy.

7°. In Greece, Babylon, India, and China:
The Theorem of Pythagoras.

8°. In the West, in Greece, Babylon, India, and China:
Pythagorean Triples.

9°. In the West, in Greece and Egypt:
Megalith Architecture.

10°. In the West and in Egypt:
Orientation of temples towards the midsummer or midwinter sunrise.

11°. In Greece and India:
Constructions of altars satisfying geometrical conditions. Wrath of Gods if constructions are not exact. Construction of a square equal in area to a given rectangle.

12°. In the West, in Egypt and India:
"Cord-Stretchers" performing geometrical constructions for ritual purposes.

13°. In Egypt, Greece, and China:
One and the same incorrect rule for the area of a circle segment.

14°. In Egypt, Babylonia, and China:
Collections of mathematical problems with solutions.

15°. In China and Egypt:
One and the same correct rule for the volume of a truncated pyramid.

In my opinion, this network of interrelations and similarities can only be explained by assuming a common origin for the mathematics and astronomy of these ancient countries.

Square Roots and Cube Roots

Chapter 4 of the "Nine Chapters" contains five problems of the following kind:

Problem 12. One has a (square) area of 55 225 (square) *pu*. What is the side of the square? Answer: 235 *pu*.

The text gives a perfectly general rule for extracting the square root. One digit after another is determined by the same method I learnt at school. The method is based on the identity

$$(19) \qquad (a+b)^2 = a^2 + b(2a+b).$$

One first chooses an approximation a such that a^2 is less than (or equal to) the given number N. Next one subtracts a^2 from N. Dividing the remainder by $2a$, one obtains a provisional value of b. Dividing the same remainder by $2a+b$, and rounding off the quotient to one significant decimal, one obtains the final value of b.

Next the square of $a+b$ is computed by means of (19) and subtracted from N, and the process can go on in the same way.

Just so, cube roots are computed by means of the formula

$$(20) \qquad (a+b)^3 = a^3 + b(3a^2 + 3ab + b^2).$$

An application is the following problem: Given the volume of a sphere, to find the diameter. The given volume is multiplied by 16 and divided by 9, and from the result the cube root is extracted. This means that the volume of a sphere is equated to

$$(21) \qquad V = \tfrac{9}{16} d^3 = \tfrac{9}{2} r^3,$$

where d is the diameter and r the radius. This is not bad, the correct value being $\frac{4\pi}{3} r^3 = 4.19\, r^3$.

Whoever found the method to extract square roots and cube roots must have been an excellent mathematician. Moreover, he must have had a decimal number system at his disposal, for his method yields one decimal after the other. In the sexagesimal number system the calculations would be much more complicated.

Note that the method to determine the next decimal consists of two steps. In a first step, the square root of $a^2 + c$ (where c is a relatively small remainder) is approximated by

$$(22) \qquad \sqrt{a^2 + c} \sim a + \frac{c}{2a}.$$

Next, in a second approximation, the numerator $2a$ is replaced by $2a+b$, where b is the correction term $c/2a$ found in the first approximation.

The approximation (22) is also found in Heron's Metrica, and in Babylonian texts. It can also be used of c is replaced by $-c$:

$$(22a) \qquad \sqrt{a^2 - c} \sim a - \frac{c}{2a}.$$

In both cases, the approximation yields a value which is slightly too large, because the square of the right hand member of (22) or (22 a) exceeds $a^2 + c$ or $a^2 - c$ by $c^2/4a^2$.

Let us now apply the formulae (22) and (22 a) to the computation of the diagonal of a unit square. A first approximation would be

$$\sqrt{1+1} \sim 1 + \tfrac{1}{2} = \tfrac{3}{2}.$$

The square of 3/2 exceeds 2 by 1/4, so the second approximation would be

$$\sqrt{(\tfrac{3}{2})^2 - \tfrac{1}{4}} \sim \tfrac{3}{2} - \tfrac{1}{12} = \tfrac{17}{12}.$$

This approximation (1;25 in the Babylonian sexagesimal notation) frequently occurs in Babylonian texts.

This square of 17/12 exceeds 2 by 1/144. So the third approximation would be

$$\sqrt{(\tfrac{17}{12})^2 - \tfrac{1}{144}} \sim \tfrac{17}{12} - \tfrac{1}{408}$$

In the Súlvasútras, this approximation is written as

$$1 + \tfrac{1}{3} + \tfrac{1}{12} - \tfrac{1}{12 \times 34}.$$

It can also be obtained by taking $1 + 1/3$ as a first approximation. The correction term $1/12$ would yield the second approximation, and the negative correction $-1/(12 \times 34)$ would yield the third approximation.

In the Babylonian text YBC 7289 (see Neugebauer and Sachs, Math. Cuneiform Texts, p. 42) the approximation

$$\text{diagonal} = 1;24,51,10$$

occurs, which is correct up to the third sexagesimal. It may have been obtained by repeated application of the formulae (22) and (22 a). It seems that the approximations (22) and (22 a) were already known in pre-Babylonian mathematics.

Sets of Linear Equations

One of the most interesting parts of the "Nine Chapters" is Chapter 8. In this chapter a systematic method is taught for solving sets of linear equations in an arbitrary number of unknowns.

Problem 1 reads in the translation of Donald B. Wagner:

The yield of 3 sheaves of superior grain, 2 sheaves of medium grain, and 1 sheaf of inferior grain is 39 *tou*.

The yield of 2 sheaves of superior grain, 3 sheaves of medium grain, and 1 sheaf of inferior grain is 34 *tou*.

The yield of 1 sheaf of superior grain, 2 sheaves of medium grain, and 3 sheaves of inferior grain is 26 *tou*.

What is the yield of superior, medium, and inferior grain?

Answer:

one sheaf of superior grain $9\frac{1}{4}$ *tou*,

one sheaf of medium grain $4\frac{1}{4}$ *tou*,

one sheaf of inferior grain $2\frac{3}{4}$ *tou*.

This problem is equivalent to a set of 3 linear equations in 3 unknowns x, y, z:

$$3x + 2y + z = 39$$

$$2x + 3y + z = 34$$

$$x + 2y + 3z = 26.$$

The method of solution is very remarkable:

Set up (with counting rods on the counting board) on the right side

superior grain 3 sheaves
medium grain 2 sheaves
inferior grain 1 sheaf
yield 39 *tou*.

In the middle and on the left set up in the same way as on the right.

Thus the pupil is required to set up what we call a matrix of 3 columns, starting with the column on the right:

```
 1   2   3
 2   3   2
 3   1   1
26  34  39
```

The column on the right comes from the first equation, the middle column from the second, and the left column from the third equation. – The text goes on:

Multiply the middle column throughout by the (number of sheaves of) superior grain on the right, and continuously subtract it.

This means: the middle column is multiplied by 3 (the number in the right upper corner), and then the right column is subtracted twice from it. The result is

```
 1   0   3
 2   5   2
 3   1   1
26  24  39.
```

Next the left column is multiplied by 3, and the right column is subtracted from it. Result

```
 0   0   3
 4   5   2
 8   1   1
39  24  39.
```

Multiplying the left column by 5 and subtracting 4 times the middle column, one obtains

$$
\begin{array}{rrr}
0 & 0 & 3 \\
0 & 5 & 2 \\
36 & 1 & 1 \\
99 & 24 & 39.
\end{array}
$$

Thus, the set of equations is transformed into the equivalent set

$$3x+2y+z=39$$
$$5y+z=24$$
$$36z=99.$$

From the last equation z can be solved, next, from the preceding equation y, and from the first equation x. All this is explained in general terms, e.g.:

As to what remains of inferior grain in the left column, the upper (number, namely 36) is the divisor, and the lower (namely 99) is the dividend for inferior grain.

That is, one has to calculate

$$z=\tfrac{99}{36}=\tfrac{11}{4}=2\tfrac{3}{4}.$$

The author of this method had a systematic mind: he knew how to explain algorithms in general terms in a clear and concise way. In his explanations, he presupposed a familiarity with the use of the counting board, just as the author of Chapter 1 did in his explanation of the Euclidean Algorithm.

In the course of matrix manipulations, negative numbers may occur. The text explains correctly how to handle these numbers.

Problems on Right-Angled Triangles

Chapter 9 of the "Nine Chapters" contains 16 problems on right-angled triangles. In the solutions, the relation

$$x^2+y^2=z^2$$

between the sides of a right-angled triangle is presupposed. The problems are of the following types:

Type 1: Given x and y, to find z.   Problems 1 and 5.
Type 2: Given y and z, to find x.   Problems 2,3,4.
Type 3: Given x and $z-y$.   Problems 6,7,8,9,10.
Type 4: Given $x-y$ and z.   Problem 11.
Type 5: Given $z-x$ and $z-y$.   Problem 12.

Type 6: Given x and $z+y$.   Problem 13.
Type 7: Given x and $y+z=\frac{7}{3}x$.   Problem 14.
Type 8: Given x and y, to find the side of the
              inscribed square.   Problem 15.
Type 9: Given x and y, to find the radius of the
              inscribed circle.   Problem 16.

In all cases, the problems are stated with definite numbers, as in Egyptian and Babylonian problem-texts, but the solutions are presented in the form of general rules. The solutions are all correct.

In all these problems, the shorter leg is called *kou*, which literally means "hook", and the longer leg *ku*, which means "thigh". The hypotenuse is called *hsien*, which means "bowstring". The units of length are:

$$1 \; chang = 10 \; ch'ih$$

$$1 \; ch'ih \;\; = 10 \; ts'un.$$

The *ch'ih* of the Han-period was about 23 cm, so one might translate *ch'ih* freely by "foot". Just so, the *ts'un* of 2.3 cm might be called "inch".

The single types are treated as follows:

Type 1. Given x and y, to find z.

This type is represented by problems 1 and 5. In Donald Wagner's translation, problem 1 reads:

9:1. The shorter leg is 3 *ch'ih*, and the longer leg is 4 *ch'ih*. What is the hypotenuse? Answer: 5 *ch'ih*.

The method of solution is presented in the form of a general rule:

Multiply the shorter leg and the longer leg each by itself, add, extract the square root. This is the hypotenuse.

This rule is equivalent to our formula

(23) $$z=\sqrt{x^2+y^2}.$$

Type 2. Given y and z, to find x.

This type is represented by problems 2, 3, and 4, every time with a different geometrical interpretation. The rule for finding the solution can be written as

(24) $$x=\sqrt{z^2-y^2}.$$

Type 3. Given $x=a$ and $z-y=d$, to find y and z.

An example is problem 6:

9:6. A pond is 1 *chang* ($=10$ *ch'ih*) square. A reed grows at its center and extends 1 *ch'ih* out of the water. If the reed is pulled to the side (of the pond), it reaches the side precisely. What are the depth of the water and the length of the reed?

Answer: The depth of the water is 1 *chang*, 2 *ch'ih*. The length of the reed is 1 *chang*, 3 *ch'ih*.

In Fig. 33 we have a right-angled triangle, in which one side $x=a=5$ and the difference $z-y=d=1$ of the other sides are given. Now

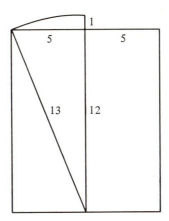

Fig. 33. The reed in the pond

$z^2 - y^2 = a^2 = 25$ is known, and since $z - y = d$ is also known, we have, by division

$$z + y = \frac{a^2}{d}.$$

Now $z + y$ and $z - y$ are known, and one finds

(25)
$$y = \frac{a^2 - d^2}{2d},$$

(26)
$$z = y + d.$$

The text gives a general rule of procedure equivalent to the formulae (25), (26). The text says: "Halve the side of the pool and multiply it by itself", etc.

The triple (5, 12, 13) is one of the simplest Pythagorean triples. The megalith builders in England and the authors of the Śulvasūtras used this triple, and it can also be obtained from the rule ascribed to Pythagoras.

Type 4. Given $x - y = d$ and $z = c$, to find x and y.

Problem 11 is of this type. The rules given in the text are equivalent to the correct formulae

$$x = s + \tfrac{1}{2}d, \quad y = s - \tfrac{1}{2}d, \quad s = \sqrt{\tfrac{1}{2}(c^2 - 2(\tfrac{d}{2})^2)}.$$

Type 5. Given $z - x = a$ and $z - y = b$, to find x, y and z.

Problem 12 is of this type:

9:12. There is a door whose height and width are unknown, and a pole whose length is unknown. (Carried) horizontally, the pole does not fit by 4 *ch'ih*. Vertically, it does not fit by 2 *ch'ih*. Slantwise, it fits exactly. What are the height, width, and diagonal of the door?

Answer:

Width   6 *ch'ih*
Height   8 *ch'ih*
Diagonal 1 *chang* (= 10 *ch'ih*).

Method: Multiply together the horizontal and vertical lacks of fit, double, extract the square root. Adding to the result the vertical lack of fit gives the width of the door. Adding the horizontal lack of fit gives the diagonal of the door.

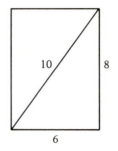

Fig. 34. The door with its diagonal

The solution given in the text is equivalent to the correct formulae

$$x = b + \sqrt{2ab}$$
$$z = a + x.$$

The problem is equivalent to a set of one quadratic and two linear equations:

$$x^2 + y^2 = z^2$$
$$z - x = a$$
$$z - y = b.$$

If x and y are solved from the second and third equation and substituted into the first, one obtains

$$(z - a)^2 + (z - b)^2 = z^2$$

or in modern notation, when the right side is reduced to zero,

$$z^2 - 2z(a + b) + a^2 + b^2 = 0.$$

The standard method to solve such a quadratic equation would be to add $2ab$ to both sides, thus obtaining

(27) $$[z - (a + b)]^2 = 2ab.$$

Extracting the square root on both sides, one obtains

(28) $$z = a + b + \sqrt{2ab}$$

which is just the solution given in the text.

The Broken Bamboo

Problem 13 is of
Type 6: Given $x = a$ and $y + z = b$, to find y.
The problem reads:

9:13. A bamboo is 1 *chang* ($= 10$ *ch'ih*) tall. It is broken, and the top touches the ground 3 *ch'ih* from the root. What is the height of the break?

Answer: $4\frac{11}{20}$ *ch'ih*.

We are given $y + z = 10$ and $x = 3$. The method of solution is:

Method: Divide the height into the product of the distance from the root by itself. Subtract the result from the height of the bamboo and halve the difference; (the result) is the height of the break.

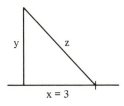

Fig. 35. The broken bamboo

This rule is equivalent to

(29) $$y = \frac{1}{2}\left(b - \frac{a^2}{b}\right).$$

Type 6 is quite similar to Type 3. In Type 3 one knows $z^2 - y^2$ and $z - y$ and one finds $y + z$ by division. In Type 6 one knows $z^2 - y^2$ and $z + y$, and $z - y$ is found by division. In both cases the identity

(30) $$z^2 - y^2 = (z - y)(z + y)$$

is presupposed.

Type 7. Given x and a linear relation

$$y + z = \tfrac{7}{3}x.$$

Problem 14, the only problem of Type 7, has been discussed already in Chapter 1. We have seen that the solution was found by first constructing a similar triangle (x',y',z') satisfying the relation

(31) $$y'+z'=\tfrac{7}{3}x'$$

and next multiplying x',y',z' by y/y' in order to obtain the given value of y (see Fig. 1 in Chapter 1).

Two Geometrical Problems

Type 8. Given a right-angled triangle with orthogonal sides x and y, to find the side s of the inscribed square.
Solution:

(32) $$s = \frac{xy}{x+y}.$$

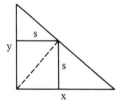

Fig. 36. Square inscribed in a triangle

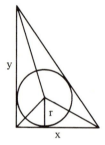

Fig. 37. Circle inscribed in a triangle

This formula may be proved as follows: A diagonal of the square divides the given triangle into two triangles. The area of the given triangle is $\tfrac{1}{2}xy$, the areas of the two partial triangles are $\tfrac{1}{2}sx$ and $\tfrac{1}{2}sy$. Hence

$$\tfrac{1}{2}xy=\tfrac{1}{2}sx+\tfrac{1}{2}sy, \quad \text{or}$$

(33) $$xy=s(x+y).$$

The same idea can also be applied to the next type.
Type 9. Given a right-angled triangle with orthogonal sides x and y, to find the diameter of the inscribed circle (Problem 16).
Solution:

(34) $$z = \sqrt{x^2+y^2}$$

(35) $$2r = \frac{2xy}{x+y+z}.$$

Proof. From the center of the circle draw lines to the three vertices. The triangle of area $\frac{1}{2}xy$ is thus divided into three triangles. Hence

$$\frac{1}{2}xy = \frac{1}{2}rx + \frac{1}{2}ry + \frac{1}{2}rz$$
$$xy = r(x+y+z).$$

Whoever invented the sequence of problems on right-angled triangles and their solutions must have been a very able mathematician. He knew the Theorem of Pythagoras, he made an ingenious use of the proportionality of sides in similar triangles, he knew how to calculate the radius of an inscribed circle in a triangle, and he knew how to solve a set of one quadratic and two linear equations.

Parallel Lines in Triangles

The last eight problems in Chapter 9 are geometrical problems of one and the same kind. To solve them, one always has to calculate the length of a line drawn in a triangle parallel to one side. The first problem of this kind reads in the translation of Wagner:

9:17. A city is 200 *pu* square. At the center of each (side) is a gate. (At a distance) 15 *pu* outside the east gate is a tree. Going out the south gate, how many *pu* does one walk to see the tree?

Answer: $666\frac{2}{3}$ *pu*.

Method: Let the divisor be the number of *pu* outside the east gate. Halve the side of the city and multiply it by itself to obtain the dividend. Dividing the dividend by the divisor gives (the answer) in *pu*.

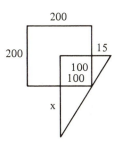

Fig. 38. The tree outside the city

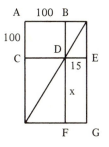

Fig. 39. Application of Euclid I,43

In Fig. 38 the unknown distance is denoted by x. We have two similar triangles, one with sides 15 and 100, and one with sides 100 and x. So we have the proportion

$$15:100 = 100:x,$$

from which the solution

$$x = \frac{100^2}{15}$$

follows. The solution can also be justified by using a theorem in Euclid's Elements (Prop. I,43), which says that in a situation like that of Fig. 39 the rectangles $ABCD$ and $DEFG$ on both sides of the diagonal are equal. I don't know whether the Chinese used Euclid's theorem or the similarity of triangles.

The other problems are similar. In problem 20 the side of a square city is unknown. Once more, one goes out of a gate until one sees a tree. To solve the problem, one has to solve a quadratic equation of the form

$$x^2 + ax = b.$$

In describing the solution of this equation, the author uses two technical terms *shih* and *tsung-fa,* which are also used in Chapter 4 in the explanation of the square-root algorithm. Most probably, the Chapters 4 and 9 are due to one and the same author. His skill in geometry and algebra is admirable.

Part B

Babylonian Mathematics

A Babylonian Problem Text

The text BM (= British Museum) 34568, published and translated by O. Neugebauer in Vol. 3 of his "Mathematische Keilschrifttexte" (Quellen und Studien Geschichte der Mathematik A 3, p. 14–22) contains 19 problems on "length, breadth and diagonal". The problems are quite similar to those of the "Nine Chapters". The equation

$$x^2 + y^2 = z^2$$

is presupposed in all solutions. The problems can be classified by types, four of which are identical with our earlier types 1,2,3 and 6:
Type 1. Given x and y, to find z.
This type is represented by Problem 1, which reads:
4 the length, 3 the breadth. What is the diagonal? The magnitude is unknown.
Two solutions are given in the text, both being valid only in the case of the (3, 4, 5)-triangle. The first solution reads: "One half of your length you

add to your breadth." This solution may be written as

$$z = x + \tfrac{1}{2} y.$$

In the case $x=3$, $y=4$ this gives the correct solution $z=5$, but not in other cases. The second solution

$$z = y + \tfrac{1}{3} x$$

also yields $z=5$. Our text is worse than the Chinese text, which gives the correct general solution

$$z = \sqrt{x^2 + y^2}.$$

I suppose the scribe who wrote BM 34568 copied the problems and solutions from an earlier text, and the correct solution of problem 1 was broken off. Clay tablets are often damaged in the course of time.

Type 2. Given y and z, to find x.
Problem 2 is of this type. The solution is correctly given as

$$x = \sqrt{z^2 - y^2}$$

for the special case $y=4$, $z=5$.

Note that problems 1 and 2 of the Babylonian text are identical with problems 1 and 3 of the Chinese text. The latter gives the solutions first for the special case of the (3,4,5)-triangle, and next formulates the general rules in words without numbers, whereas the Babylonian text only treats the special case (3,4,5).

Type 3. Given $x=a$ and $z-y=d$, to find y and z.
This type is represented by the very remarkable Problem 12 of the Babylonian text. It reads:

A reed stands against a wall. If I go down 3 yards (at the top), the (lower) end slides away 9 yards. How long is the reed, how high is the wall?

Here we are given $x=9$ and $z-y=3$. The solution is

$$z = \frac{1}{2} \frac{9^2 + 3^2}{3} = 15$$
$$y = \sqrt{z^2 - x^2} = 12.$$

The solution is correct, but the extraction of the square root is not necessary. Much simpler:

$$y = z - d = 15 - 3 = 12.$$

The Chinese solution of the same type of problem is essentially the same, but the extraction of the square root does not occur. Problem 8 of Chapter 9 of the "Nine Chapters" reads:

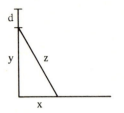

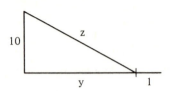

Fig. 40. The reed against the wall　　　　　　Fig. 41. The timber leaned against the wall

9:8. The height of a wall is 1 *chang* (= 10 *ch'ih*). A timber is leaned against the wall so that its top is precisely at (the top of) the wall. If the timber is pulled 1 *ch'ih* away from (the wall), the timber reaches the ground. What is the length of the timber?

Answer: 5 *chang*, 5 *ch'ih* (= 55 *ch'ih*).

Method: Multiply the height of the wall, 10 *ch'ih*, by itself, and divide by the number of *ch'ih* pulled away. Add to the result the number of *ch'ih* pulled away, and halve the sum. (The result) is the length of the timber.

In this case we are given

$$x = a = 10, \quad z - y = d = 1,$$

and z is calculated according to the formula

(36) $$z = \frac{1}{2} \left(\frac{a^2}{d} + d \right),$$

which agrees with the Babylonian rule

(37) $$z = \frac{1}{2} \frac{a^2 + d^2}{d}.$$

The text BM 34568, in which the rule (37) is used, is from the Seleucid period, which began in 311 B.C. with the accession of the Greek king Seleukos I to the Babylonian throne. However, the geometrical picture associated with Problem 12 of the Seleucid text is quite similar to that of Problem 9 of the Old-Babylonian text BM 85196. In a free translation this problem reads:

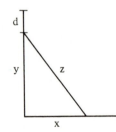

Fig. 42. The beam standing against the wall

A patû (beam?) of length 0;30 (stands against a wall). The upper end has slipped down a distance 0;6. How far did the lower end move?

In this case we have

$$z = 0;30$$

$$y = 0;30 - 0;6 = 0;24$$

hence, in full accordance with the text,

$$x = \sqrt{z^2 - y^2} = 0;18.$$

In this case the problem is of type 2, but the geometrical picture associated with it is the same as in Problem 9 of the Seleucid text BM 34568. Obviously, the early and the late text are closely related. Neugebauer too stresses the continuity of the tradition during more than 1000 years (Mathematische Keilschrifttexte, Vol. 3, p. 22).

The next problem of the Seleucid text belongs to

Type 6. Given $x = a$ and $y + z = b$, to find y and z.

The text deals with the case $a = 3$, $b = 9$, and gives the correct solution

$$y = \tfrac{1}{2}(b^2 - a^2)/b = 4$$

$$z = b - y = 5.$$

Problems 5 and 7 are again of type 1. In problems 6 and 8, x and y are given and the area F of the rectangle is calculated as $F = xy$.

Problems 9–18 belong to the following new types:

Type 10 (Problem 9). Given $x + y = s$ and $xy = F$, to find x and y.

Type 11 (Problem 15). Given $x - y = d$ and $xy = F$, to find x and y.

In problem 9, we are to find two numbers whose sum and product are given, and in problem 15 the difference and the product are given. These two problems belong to the standard problems of Babylonian algebra. Their solution will be discussed later in a section entitled "The Sum and Difference Method".

Most interesting is the next type:

Type 13 (Problems 14, 17 and 18). Given $x + y + z = s$ and $xy = F$, to find z.

The calculations given in the text are equivalent to the correct formula

$$z = \tfrac{1}{2}(s^2 - 2F)/s.$$

It is very remarkable that the solution of problem 18 is presented as a general rule, as follows:

Multiply (the sum of) length, breadth and diagonal by itself. Multiply the area by 2. You subtract this product from the (square of the sum of length, breadth and) diagonal. The difference multiply by $\tfrac{1}{2}$. By what factor should you multiply (the sum of) length, breadth and diagonal? The diagonal is this factor.

This is the only case in our Babylonian text in which a general rule is formulated in words without numbers. The rule is quite similar to the general rules in the Chinese text. The Chinese text is more satisfactory from a mathematician's point of view, because it gives general rules in *all* cases.

The similarities between the Chinese and the Babylonian text are striking. The conclusion that they were derived from a common source seems unavoidable.

Quadratic Equations in Babylonian Texts

We have seen that the problems 9:12 and 9:20 in the "Nine Chapters" led to quadratic equations of the forms

$$x^2 + b = ax$$

and

$$x^2 + ax = b,$$

and that these equations were solved correctly.

The Babylonians too were able to solve quadratic equations. The Old-Babylonian text BM 13901, published and translated by O. Neugebauer in Vol. 3 of his standard edition "Mathematische Keilschrifttexte" (Quellen und Studien Gesch. Math. A 3, p. 1–14) contains the following sequence of quadratic equations:

1. $$x^2 + x = 0;45$$
2. $$x^2 - x = 14,30$$
3. $$x^2 - \tfrac{1}{3}x^2 + \tfrac{1}{3}x = 0;20$$
4. $$x^2 - \tfrac{1}{3}x^2 + x = 4,46;40$$
5. $$x^2 + x + \tfrac{1}{3}x = 0;55$$
6. $$x^2 + \tfrac{2}{3}x = 0;35$$
7. $$11x^2 + 7x = 6;15$$

For instance, problem 2 reads:
I have subtracted (the side of) the square from the area, and it is 14,30.
The number 14,30 signifies

$$14 \times 60 + 30 = 870.$$

So the problem is, to solve the equation

$$x^2 - x = 870.$$

The solution is given in words:
Take 1, the coefficient (of the side of the square). Divide 1 into two parts.
$0;30 \times 0;30 = 0;15$ you add to 14,30.
14,30;15 has the square root 29;30.
You add to 29;30 the 0;30 which you have multiplied by itself, and 30 is (the side of) the square.

At school we learnt that one solution of the equation

$$x^2 - ax = b$$

is

$$x = \tfrac{1}{2}a + \sqrt{(\tfrac{1}{2}a)^2 + b},$$

and this is just what our text teaches.

The Method of Elimination

Problem 14 of the same text BM 13901 leads to a set of two equations with two unknowns:

$$x^2 + y^2 = 25,25$$
$$y = \tfrac{2}{3}x + 5.$$

If y is solved from the second equation and substituted into the first, one obtains a quadratic equation

$$ax^2 + 2bx = c$$

for x, in which,

$$a = 1 + 0;40^2 \ = 1;26,40$$
$$b = 5 \times 0;40 \ = 3;20$$
$$c = 25,25 - 5^2 = 25, \ 0$$

The text first calculates the coefficients a,b,c. Next the quadratic equation for x is solved by the standard method explained earlier in the same text, and finally y is calculated as $\tfrac{2}{3}x + 5$.

Problem 18 of the same text is equivalent to 3 equations with 3 unknowns:

$$x^2 + y^2 + z^2 = 23,20$$
$$x - y = 10$$
$$y - z = 10.$$

The method of solution is again the same: x and y are expressed in terms of z, and a quadratic equation is obtained for z, which is solved by the standard method.

This problem 18 is similar to the Chinese problem 12. In both cases the differences $x-y$ and $y-z$ (or $z-x$ and $z-y$) are given, and a quadratic equation involving x^2+y^2 and z is proposed. In both cases x and y are expressed in terms of z, and a quadratic equation for z is obtained, which is solved by the standard method.

In the "Arithmetica" of Diophantos, and also in the "Algebra" of Al-Khwārizmī, sets of equations in several unknowns are always reduced to single equations in one unknown. There are several methods to achieve this aim. One of the methods is the "Method of Elimination": all unknowns but one are eliminated by substitution. We have seen that this method was used by the Babylonian mathematicians as well as by their Chinese colleagues.

More interesting is another method, frequently used by Diophantos and also by the Babylonians, namely:

The "Sum and Difference" Method

When Diophantos of Alexandria wants to find two numbers x and y, of which the sum a is given, and who also have to satisfy another condition, he frequently puts

(38)
$$x = \tfrac{1}{2}a + t$$
$$y = \tfrac{1}{2}a - t$$

and thus reduces the problem to a problem with one unknown t. Just so, when the difference $x-y=d$ is given, he sometimes puts

(39)
$$x = t + \tfrac{1}{2}d$$
$$y = t - \tfrac{1}{2}d.$$

The Babylonians often applied the same method. For instance, in the text BM 13 901, Problem 8, the set of equations

(40)
$$x^2 + y^2 = S = 21,40$$
$$x + y = a = 50$$

is proposed (see Neugebauer, MKT III, p. 7). If we put

$$x = \tfrac{1}{2}a + t$$
$$y = \tfrac{1}{2}a - t$$

we find for t the equation

$$2(\tfrac{1}{2}a)^2 + 2t^2 = S$$

and the solution

$$t = \sqrt{\tfrac{1}{2}S - (\tfrac{1}{2}a)^2}$$
$$x = \tfrac{1}{2}a + t$$
$$y = \tfrac{1}{2}a - t$$

which is just the solution given in the text.
In problem 9 of the same text, we have

(41)
$$x^2 + y^2 = S = 21{,}40$$
$$x - y = d = 10 .$$

The solution is given as

$$t = \sqrt{\tfrac{1}{2}S - (\tfrac{1}{2}d)^2}$$
$$x = t + \tfrac{1}{2}d$$
$$y = t - \tfrac{1}{2}d .$$

By the same method, sets of equations of the form

(42)
$$xy = F$$
$$x + y = a$$

or

(43)
$$xy = F$$
$$x - y = d$$

were solved by the Babylonians.

More complicated problems were sometimes reduced to the standard forms (42) and (43). For a typical example see my book "Science Awakening I", p. 63–65.

In Chapter 3 we shall see that the same standard forms (40), (41), (42), and (43) also occur, in geometrical wording, in Euclid's Elements and Data, and that Euclid's method of solution is again the "Sum and Difference" method.

For more information about Babylonian algebra I may refer to the paper of H. Goetsch: Die Algebra der Babylonier, Archive for History of Exact Sciences 5, p. 79–153.

Part C

General Conclusions

Chinese and Babylonian Algebra Compared

When speaking of "Babylonian Algebra" we need not distinguish between early and late Babylonian texts. The late and early problem texts are of just the same kind; they use the same technical expressions and the same number system. One and the same tradition was preserved from about 1700 to about 100 B.C., without any trace of Greek influence, despite the fact that the later texts were written in the Hellenistic period.

Between Babylonian and Chinese algebra there are many similarities. An essential element in both kinds of algebra is the solution of quadratic equations. In China as well as in Babylon, sets of linear equations were solved by eliminating one unknown after the other. Numerical methods for calculating square roots and cube roots were known in both countries.

In Chapter 1 we have seen that the Babylonian method of computing Pythagorean triples is equivalent to the Chinese method, and in Chapter 2 we have seen that there is a striking simirarity between the problems on right-angled triangles in Chinese and Babylonian texts. These similarities all point towards a common origin.

The question now arises: Is it possible that Chinese algebra was derived from Babylonian algebra?

For several reasons, I don't think that such a derivation is possible.

First, Chinese algebra is on a higher mathematical level. The "Euclidean Algorithm" is found in the Chinese "Nine Chapters", but not in any one of our numerous Babylonian texts. The Chinese text presents the methods of solution as general rules, whereas the Babylonians formulate general rules only in a few exceptional cases. The Chinese text has a systematic matrix method for solving sets of linear equations in a arbitrary number of unknowns, whereas in Babylonian texts no such general method is explained.

Secondly, the Chinese text is richer in geometrical notions and methods. In a Chinese astronomical text we find a proof of the Theorem of Pythagoras. In Chapter 9 of the "Nine Chapters" we find two problems on the inscribed square and circle of a triangle and several problems on persons just seeing a tree outside the walls of a city. The Chinese have the idea of proportionality of sides in similar triangles. All this is missing in our Babylonian texts.

For a scribe who translates a set of problems into another language it is easy to translate words such as "add", "subtract", "multiply", and "divide", but the translation of geometrical terms and of expressions like "a reed extends 1 *ch'ih* out of the water" is much more difficult. Therefore, it is only natural for translators and copyists to leave out problems in which such expressions occur, and to restrict themselves to purely algebraic problems and solutions. The inverse process, namely the addition of geometrical problems by a scribe is highly improbable.

A decisive argument against the hypothesis of a Babylonian origin of Chinese algebra is the system of calculation used in the Chinese texts. The Chinese use the decimal system for integers, and they use mixed fractions m/n. The Babylonians use the sexagesimal system for integers as well as for fractions. In the Chinese system, the division of 12 by 5 yields $2\frac{2}{5}$, just as we have learned at school, but if a Babylonian scribe has to divide 12 by 5, he looks up $5^{-1} = 0;12$ in a table of reciprocals, and multiplies the result by 12. A completely different method.

Now, if a scribe finds in his source a calculation making use of common fractions m/n, he can easily make the same calculation in the sexagesimal system. He only has to replace every fraction m/n by the product mn^{-1}. If n is not a regular sexagesimal number, he can use an approximation. The inverse process, namely the passage from sexagesimal fractions to common fractions, is much more difficult and inconvenient. Let me illustrate this by an example taken from Ptolemy's Almagest I 12.

According to Ptolemy, Eratosthenes found that the arc on the meridian between the highest and the lowest culmination point of the sun is

$$\tfrac{11}{83} \cdot 360°.$$

Manitius, the German translator of the Almagest, had no difficulty to convert this figure into degrees, minutes, and seconds. His result is

$$47°\,42'\,40''$$

(39″ would be more exact, but this does not matter). Ptolemy too was able to perform this conversion, for he says that the arc lies between

$$47°\,40' \quad \text{and} \quad 47°\,45'.$$

Now let us try to perform the inverse process, namely to find a common fraction m/n approximating

$$\frac{47°\,42'\,40''}{360°} = \frac{171\,760''}{1\,296\,000''} = \frac{2\,147}{16\,200}.$$

Applying to the pair of numbers 2 147 and 16 200 the Euclidean Algorithm, one obtains

$$16\,200 = 7 \times 2\,147 + 1\,171$$
$$2\,147 = 1 \times 1\,171 + 976$$
$$1\,171 = 1 \times 976 + 195$$
$$976 = 5 \times 195 + 1.$$

Now, in order to obtain an approximation, one may neglect the last remainder 1, and replace the numbers 976 and 195 by 5 and 1:

$$5 = 5 \times 1 + 0.$$

Going upwards in the chain of divisions, retaining the quotients 1, 1, 7, one obtains

$$1 \times \ \ 5 + 1 = \ \ 6$$
$$1 \times \ \ 6 + 5 = 11$$
$$7 \times 11 + 6 = 83.$$

It follows that 11/83 is a good approximation of the original fraction 2 147/16 200.

One sees that the conversion of a fraction m/n into a sexagesimal fraction, accurate to seconds, is an easy operation, which can be performed by a well-instructed scribe, but that the inverse procedure is very cumbersome. A good mathematician like Eratosthenes can apply the Euclidean algorithm and obtain a reasonable result, but a trained scribe cannot.

We have seen that the Babylonian method to calculate square roots can be derived from the Chinese method, but not conversely. The Chinese method for finding the next decimal consists of two steps. The first step is based on the approximation

$$\sqrt{a^2 + c} \sim a + \frac{c}{2a}$$

which is also used in Babylonian texts. The second step is closely connected with the decimal number system: this step is not found in Babylonian texts.

For these reasons, it would be unreasonable to assume that Chinese algebra comes from a Babylonian source. Also, a Sumerian source must be excluded, because the Sumerian number system too was sexagesimal. The only remaining possibility is a *pre-Babylonian* common source of Chinese and Babylonian algebra.

The Historical Development

In Chapter 1, we have seen that methods for calculating Pythagorean triples were known in Neolithic Western Europe, in Greece, in Babylonia, in India, and in China, and that in all these countries the methods were based on the Theorem of Pythagoras and on the identity

$$(a+b)(a-b) = a^2 - b^2.$$

We have also seen that the same ideas about the ritual importance of geometrical constructions are found in Greek and in Sanskrit texts. Greek geometers and Hindu ritualists used the same geometrical construction for obtaining a square equal in area to a given rectangle. Geometrical proofs of the same kind, based on the addition and subtraction of plane areas, were given in the Śulvasūtras, in Euclid's Elements, and in a Chinese astronomical text.

In Chapter 2, we have seen that Chinese and Babylonian mathematics are closely connected. Areas of plane figures and volumes of solids were calculated by the same rules in Babylonia and in China. The correct rule for calculating the volume of a truncated pyramid was known in China as well as in Egypt. Even more significant is an incorrect rule for the area of a circle segment, which is found in an Egyptian papyrus, in the Chinese "Nine Chapters", and in Heron's Metrica. Quadratic equations and sets of linear and quadratic equations in several unknowns were solved by essentially the same methods in Babylonia and China.

In my opinion, the only reasonable explanation of these similarities is the assumption of a common origin. In neolithic Europe, in the domain of Indo-European languages, an oral tradition of arithmetic, algebra, and geometry must have existed, with applications to architecture, astronomy, and ritual.

It seems that the most faithful reflection of this oral tradition is found in the Chinese "Nine Chapters". Here we find a complete explanation of calculations in the decimal system, including the simplification of fractions m/n by means of the Euclidean algorithm. This algorithm was known in China and in Greece, but no trace of it is found in any one of our numerous Babylonian and Egyptian texts. In the "Nine Chapters" we also find a systematic method for extracting square roots and cube roots, and a completely clear exposition of the matrix method for solving sets of linear equations in an arbitrary number of unknowns. In these respects the Chinese text is unique.

The ninth chapter of the "Nine Chapters" is very much similar to Babylonian problem texts, but it contains several purely geometrical problems which are not found in our cuneiform texts. Also, the Chinese text always presents general rules, whereas the Babylonian texts usually give numerical examples only. Most probably, the Chinese text and the Babylonian text BM 34568 were derived from a common prototype, which was much more geometrical and much more systematic than the totality of our Babylonian texts.

In the domain of geometry, the "Nine Chapters" are far superior to the Babylonian texts. The correct rules for calculating the volumes of pyramids, cones, truncated pyramids, and truncated cones are found in the Chinese text only. The rule for the truncated pyramid appears in an Egyptian papyrus, but the Babylonians calculated the volumes of truncated cones and pyramids by incorrect formulae. The inscribed square and circle of a right-angled triangle appear in our Chinese text only. Similar triangles occur in the "Nine Chapters", but not, as far as I know, in any Babylonian text.

The inevitable conclusion from the facts just mentioned seems to be: All texts discussed in the present chapter and in Chapter 1 are ultimately derived from one consistent mathematical theory, which is most faithfully reflected in the Chinese texts of the Han-period.

Two lines of transmission of this theory may be distinguished. In the Chinese and Babylonian texts and in Middle Kingdom Egyptian papyri we find *sets of problems with solutions.* In the present chapter I have restricted myself to such problem texts, and I have shown that they are very much similar in China, Babylon, and Egypt. On the other hand, in Chapter 1 I have collected the traces of an oral tradition of *geometrical constructions,* a tradition which existed already in the Neolithic Age in Western Europe, and which was continued by the Egyptian cord-stretchers and the Hindu ritualists.

In both traditions, the Theorem of Pythagoras played a central role. Also, methods of calculating Pythagorean triples formed a part of both traditions.

We have seen that there are good reasons to assume that these triples were not found empirically, but computed systematically by the rules explained at the beginning of Chapter 1, rules which are based on the Theorem of Pythagoras. If this is assumed, the development of the two traditions must have been, in a rough schematic simplification, as follows:

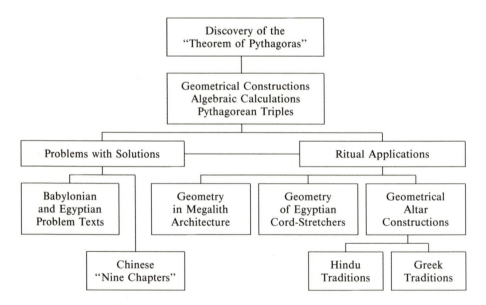

The development indicated on the four upper levels of this schema must have taken place between (say) 4000 and 1500 B.C. The three boxes at the bottom refer to extant texts written between 900 B.C. and A.D. 220.

Of course, our schema is too simple. Several important discoveries such as the Euclidean algorithm, the computation of square roots, the systematic solution of quadratic equations and sets of linear equations, the rules for calculating areas and volumes, are not included. We cannot include these discovereis, because their history is unknown.

In the case of the Theorem of Pythagoras the situation is slightly more favourable. We have the Old-Babylonian text Plimpton 322 (see Chapter 1), in which a long list of Pythagorean triples occurs. The list was composed by someone who knew the Theorem of Pythagoras. Also, the geometrical constructions in the Śulvasūtras are based on the Theorem of Pythagoras. So we are justified in assuming that the discovery of the theorem preceded the geometrical constructions and the calculation of Pythagorean triples, as indicated in the schema.

From the third level downwards, we may distinguish two lines of development. On the right side of the schema, we have the oral tradition of the cord-stretchers, who were experts in geometrical constructions with ritual applications. They were active in the British isles, in Egypt and in India. From their activity in Egypt we know very little, the most important testimony being that of Democritos, who says that the "harpedonaptai" were experts in "constructing lines with proofs", but from India we have the Brāhmanas and Śulvasūtras, in which altar constructions by means of stretched cords are described in great detail, *with proofs*. From the close connection between Greek and Hindu traditions concerning geometrical altar constructions, we may conclude that the development of these constructions took place in the realm of Indo-European languages. I suppose that oral traditions existed long before the rules were written down in the Śulvasūtras.

On the left side of the schema, I have sketched the development of a school tradition: a method of teaching mathematics by means of problems with solutions. The method is known from Babylonian and Egyptian texts written between 2050 and 1530 B.C., but by far the best and most extensive set of problems is contained in the Chinese "Nine Chapters".

Chapter 3

Greek Algebra

What is Algebra?

Our word Algebra is derived from the Arabic expression
al-jabr wa'l muqābala
which occurs in the title of the first Arabic textbook on Algebra, written by
Al-Khwārizmī in the ninth century A.D. and dedicated to the Caliph Al-
Ma'mūn. The two words *jabr* and *muqābala* denote two simple operations
necessary for reducing linear and quadratic equations to their normal
forms. *Al-jabr,* which may be translated as "restoration" or "completion",
is the addition of one and the same term to both sides of an equation in order to eliminate negative terms. *Al-muqābala,* which we may translate as
"reduction" or "balancing", is the cancelling of terms or parts of terms that
occur on both sides. Thus, the equation

$$50 + x^2 = 29 + 10x$$

which occurs in Rosen's translation of Al-Khwārizmī's Algebra on page 40,
is reduced by *al-muqābala* to

$$21 + x^2 = 10x$$

which in Rosen's translation reads:

"There remains twenty-one and a square, equal to ten things."

Thus it appears that Algebra, in the sense of Al-Khwārizmī, is just *the
art of reducing and solving equations.*

Later Arabic authors use the word Algebra (al-jabr) in the same sense.
For instance, the famous poet Omar Khayyām, who was also a great
mathematician, writes:

One of the mathematical theories needed in that part of the philosophical sciences which
is known as "mathematical sciences" is *the art of algebra,* which aims at the determination of
unknown numerical or geometrical quantities (translated from the French translation of F.
Woepcke: L'algèbre d'Omar Alkhayyâmî, Paris 1851).

In the Renaissance, Algebra was still considered as an art. The title of
Cardano's famous book, which deals among other things with the solution
of equations of degrees 3 and 4, was "Ars magna", which means "Great
Art".

In modern algebra, the emphasis has shifted from the solution of equations to the investigation of the structure of groups, fields, etc., but still the solution of equations and sets of equations is an essential and historically important part of what we call Algebra. After all, Modern Algebra took its starting point from the work of Galois on the possibility of solving equations by means of radicals!

The Role of Geometry in Elementary Algebra

For solving equations, one needs algebraic identities like

(1)
$$(a+b)^2 = a^2 + b^2 + 2ab.$$

This identity appears, in geometrical form, in Euclid's Elements as Proposition II,4:

If a straight line is cut at random, the square on the whole is equal to the squares on the segments and twice the rectangle contained by the segments.

Al-Khwārizmī uses the identity (1) for solving equations of the form

(2)
$$x^2 + ax = b$$

by adding $(\tfrac{1}{2} a)^2$ to both sides and extracting the square root, thus obtaining

(3)
$$x + \tfrac{1}{2} a = \sqrt{(\tfrac{1}{2} a)^2 + b}.$$

Al-Khwārizmī gives two geometrical proofs of the identity (1). The diagram accompanying his second proof is reproduced in our Fig. 43, side by side with Euclid's diagram illustrating his proof of II,4.

Cardano too proves algebraic identities by means of diagrams. In his "Ars magna", the solution of biquadratic equations is based on the identity

(4)
$$(a+b+c)^2 = (a+b)^2 + 2ac + 2bc + c^2,$$

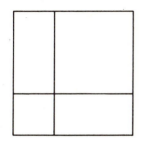

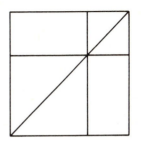

Fig. 43. Diagram from the Algebra of Al-Khwārizmī, p. 16 in Rosen's translation

Fig. 44. Diagram to Euclid, Prop. II,4

which he proves by drawing a diagram of a square divided into two squares and four rectangles (see Fig. 45).

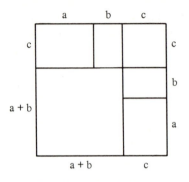

Fig. 45. Cardano's diagram

From the very beginning, algebra has always been closely connected with geometry. In Babylonian problem texts, the unknown quantities are very often called "length" and "width", and their product "area". The product of a number by itself is called "square", the number itself "side" (of the square). In problems with three unknowns, these are sometimes called "length", "width", and "height", and their product "volume".

In Greek arithmetics too, the product of a number by itself is called "tetragonon", i.e. square. This term is found in all texts from Plato to Diophantos. Numbers of the form $m \cdot n$ with $m \neq n$ were called "oblong numbers", which means that they were visualized as rectangles.

Three Kinds of Algebra

The Babylonians do not hesitate to add an area to a line segment. Thus, the old-Babylonian problem text AO 8862 (see my Science Awakening I, p. 63) begins as follows:

Length, width. I have multiplied length and width, thus obtaining the area. Next I added to the area the excess of the length over the width: (result) 183.

From this wording we see that for the Babylonians the length, width, area etc. were mainly considered as *numbers,* which can be added and multiplied without any restriction.

The Greeks, on the other hand, never add line segments to areas. They made a clear distinction between *numbers* and *geometrical quantities* (line segments, areas, and volumes). "Numbers" in the Greek sense are only rational numbers m/n. Plato and Euclid don't even allow fractions, but Eratosthenes and Archimedes used fractions like 11/83 and 22/7, and in the "Arithmetika" of Diophantos all positive fractions m/n are admitted as solutions of equations.

For the Greeks, irrational numbers don't exist. In Book 6 of his "Arithmetika", Diophantos considers the equation

$$6x^2 + 3x = 7$$

and says: "If the equation were soluble, the square of half the coefficient of x, together with the product of the coefficient of x^2 and the number 7, should be a square." Obviously, Diophantos was familiar with the numerical solution of quadratic equations, and he knew that a rational solution is only possible if the determinant is a square.

In the special case of a pure quadratic equation

(5) $$x^2 = C$$

the solution in numbers, in the Greek sense of the word "number", is possible only if the given integer C is a square. On the other hand, if C is a polygonal area, the geometrical solution of (5) is always possible. Every polygon can be transformed into a rectangle and next into a square of equal area by a geometrical construction known to Euclid as well as to the authors of the Śulvasūtras. So, if one wants to solve quadratic equations in full generality, one has to consider the unknown x and the coefficients of the equations as *geometrical quantities*.

Omar Khayyām makes a clear distinction between problems in which the unknown is a number and problems in which it is a measurable magnitude ("une grandeur mesurable" in Woepcke's translation). He distinguishes four kinds of continuous magnitudes: the *line,* the *surface,* the *solid,* and the *time.* He reminds us that Euclid first proves certain theorems concerning proportionality of geometrical magnitudes, and next, in his seventh book, presents demonstrations of the same theorems for numbers (Omar Khayyām Algèbre, p. 5–6 in the translation of Woepcke).

Thus we see that Omar Khayyām distinguishes two branches of algebra, namely *Numerical Algebra,* in which the unknowns are numbers to be calculated, and *Geometric Algebra,* in which the unknowns are geometrical magnitudes to be constructed.

The algebra of Omar Khayyām is mainly geometric. He shows that cubic equations, in which the unknowns are line segments, can be solved by means of intersections of conic sections.

Omar Khayyām's work on cubic equations is a direct continuation of early Greek investigations. Hippocrates of Chios had shown that the problem of "doubling the cube" can be reduced to the construction of two mean proportionals between two given line segments a and b:

(6) $$a : x = x : y = y : b.$$

Archytas, Eudoxos, and Menaichmos have given geometrical solutions of this algebraic problem. Menaichmos solved it by means of an intersection of two conic sections (see my Science Awakening I, p. 162). Omar Khayyām himself mentions the Elements and Data of Euclid, in which quadratic equations and sets of linear and quadratic equations are solved by means of geometrical constructions.

Omar Khayyām knows very well that earlier authors sometimes equated geometrical magnitudes with numbers. He avoids this logical inconsistency by a trick, introducing a unit of length. On p. 14 he writes:

Every time we shall say in this book "a number is equal to a rectangle", we shall understand by the "number" a rectangle of which one side is unity, and the other a line equal in measure to the given number, in such a way that each of the parts by which it is measured is equal to the side which we have taken as unity.

The same trick, namely in the introduction of a fixed unit of length e, was used by Réné Descartes in his "Géométrie" for defining the product of two line segments a and b as a line segment c. He defined:

$$ab = c \quad \text{means} \quad e : a = b : c.$$

Essentially the same artifice was also used by the author of Book 10 of Euclid's Elements. In this book, which deals with irrational line segments and areas, a fixed line segment e is introduced, which is called "the rational", or in a more literal translation "the expressible" (ἡ ῥητή). Line segments are called rational (ῥητός) if their square is commensurable with the square on e. All other line segments are called irrational (ἄλογος). One of the main objects of Book 10 is a classification of irrational lines according to algebraic criteria. For an analysis of the principal ideas of Book 10 I refer to my Science Awakening, p. 168–172.

If we now compare the Babylonian texts with Euclid, Al-Khwārizmī, and Omar Khayyām, we may distinguish three kinds of algebra:

A. *Mixed Algebra* of the Babylonian type, in which line segments and areas are added together and equated to numbers,

B. *Numerical Algebra,* in which only rational numbers m/n are admitted as coefficients and solutions of equations, as in the "Arithmetika" of Diophantos,

C. *Geometric Algebra* like the algebra of Omar Khayyām, in which line segments, areas, and volumes are strictly kept apart. We shall see presently that in Book 2 of Euclid's Elements the fundaments of this kind of algebra are laid.

On Units of Length, Area, and Volume

Our distinction between three types of algebra was based on Babylonian, Greek, and Arabic texts. The question now arises: Does the hypothetical Pre-Babylonian Algebra, which I have tried to reconstruct in Chapters 1 and 2, fit into this classification? We shall see that it does not: the situation is slightly more complicated.

As we have seen, the most faithful exposition of pre-Babylonian mathematics is the Chinese text entitled "Nine Chapters of the Mathematical Art". In this text, line segments and numbers are never added to areas or volumes. Line segments are usually expressed as rational multiples of a standard unit, the *ch'ih:* a small "foot" of approximately 23 cm. Another unit, which is mainly used in the calculation of areas, is the *pu* of approximately 1 1/3 m. The main unit of area is a square *pu;* this unit is also called *pu.* The main unit of volume is a cubic *ch'ih;* this unit is also called *ch'ih,* and 240 of these units are 1 *mou.* Numbers, line segments, areas, and volumes are strictly separated, and the units of length, area, and volume are always explicitly mentioned.

In the Śulvasūtras the fundamental unit of length is the *purusha,* the height of a man with his arms stretched upwards. The same word *purusha* is also used for the square *purusha,* the fundamental unit of area in the altar geometry of the Śulvasūtras. As in the Nine Chapters, one and the same word is used for a unit of length and the corresponding unit of area.

The same usage can be found in Greek texts. Plato, in his dialogue Theaitetos, considers squares having an area of three or five square feet and calls them "tripodos" and "pentapodos", which means "of three feet" and "of five feet", thus using the same word foot to denote a unit of area as well as a unit of length.

The "foot" is, of course, a popular unit of length, but the square foot is not a popular unit of area. No peasant would ever express the area of his property in square feet. The speaker in the dialogue is the geometer Theaitetos, who gives an account of a lecture delivered by the geometer Theodoros. Most probably the use of a square foot as a unit of area is not derived from popular usage, but from the language of early Greek geometers like Theodoros and Theaitetos.

The fact that Plato's geometers as well as the Hindu and Chinese geometers denote their unit of area by the same word as the corresponding unit of length seems to indicate that this usage was derived from pre-Babylonian geometry.

This conclusion is confirmed by the Babylonian text just quoted. In fact, when the Babylonian scribe says "I have multiplied length and width, thus obtaining the area", he implicitly supposes that the unit of area is a square having a side of one unit of length. In the Nine Chapters and in the Śulvasūtras, this implicit assumption is made explicit.

Greek "Geometric Algebra"

The term "Geometric Algebra" was coined by the Danish mathematician H. G. Zeuthen, the author of an excellent book "Die Lehre von den Kegelschnitten im Altertum". Zeuthen noted that in the "Konika" of the great Greek geometer Apollonios of Perge, the characteristic properties of conic sections were formulated by means of geometrical operations on line segments on the one hand, and plane areas on the other, which have the same properties as the addition and multiplication of real numbers in our school algebra. In Euclid's Elements as well as in Apollonios' Konika, line elements are added to and subtracted from each other, and similarly with plane areas, and two line elements are combined to form a rectangular area. The principles of this "geometric algebra" will now be explained in greater detail.

In plane geometric algebra, as it is used by Euclid and Apollonios, the fundamental relation between line segments or between areas is *equality*. Two polygons are called equal if they are equal in area. In Euclid's Elements, the notion equality remains undefined, but it is subjected to axioms called "common notions" of the following kind:

Axiom 1. Things which are equal to the same thing are equal to one another.

Axiom 2. If equals be added to equals, the wholes are equal.

Axiom 3. If equals be subtracted from equals, the remainders are equal.

In plane geometric algebra there are three fundamental operations, which can be defined as follows:

1. The *sum* of two line segments a and b is a line segment c which can be divided into two parts a' and b' that are equal to a and b respectively. Modern notation $a+b=c$.

Fig. 46. Definition of $a+b$

Note that the sum is unique in the sense of equality because of Axiom 2.

2. The *sum* of two polygons A and B is a polygon C which can be divided into two parts A' and B' that are equal to A and B. Modern notation $A+B=C$. Once more, the sum is unique because of Axiom 2.

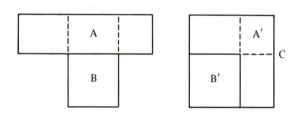

Fig. 47. Example of $A+B=C$

The example given in Fig. 47 is drawn from the proof of Proposition II,6 in Euclid's Elements. Euclid proves that a rectangle A and a square B are together equal to a square C.

3. The *"product"* of two line segments a and b is a rectangle R contained by two line segments a' and b' that are equal to a and b. Modern notation $ab=R$. This "product" is a geometrical object, not the result of a numerical multiplication.

Fig. 48. Definition of $ab = R$

In Euclid's text, the expression "the rectangle contained by a and b" is often applied to cases in which the line segments a and b are not perpendicular to each other. Therefore, in order to form a rectangle, a and b must be replaced by equal line segments a' and b', which are perpendicular and span a rectangle. For instance, in Fig. 44, the line segments a and b lie in one line. It follows that "the rectangle" on a and b" is not a definite figure situated in the plane: it is quantity that can be added to other similar quantities or equated to other figures of quite different shapes. In other words, the "rectangle" is just an object of algebraic operations.

In space we have two more operations: the addition of volumes and the "multiplication" of a line segment with a plane polygon by forming a prism with base equal to the given polygon and height equal to the given line segment.

The inverse operation of addition is the subtraction of a smaller line segment or area or volume from a larger one. Axiom 3 ensures that the result is unique in the sense of equality.

In a recent paper entitled "On the Need to Rewrite the History of Greek Mathematics", Archive for History of Science 15, p. 67–114, Sabetai Unguru flatly denies the existence of Geometric Algebra in the pre-Christian era. In my answer "Defence of a Shocking Point of View" in the same Archive, Vol. 15, p. 199–210, I have shown that Unguru's characterization of "algebraic thinking" is unhistorical, and that Geometric Algebra, as we find it in the work of Euclid, is a reality. The present investigation, which shows that the Algebra of Omar Khayyām is a natural continuation of the work of Menaichmos and Euclid, confirms this conclusion.

However, I must admit that there is an element of truth in Unguru's criticism. In my book "Science Awakening", I had considered Greek geometric algebra just as a translation of Babylonian algebra into the language of geometry. I now see that this view is only partly correct, and that Greek geometric algebra was a result of a synthesis between earlier geometrical traditions and Babylonian algebra. This modified view will be explained in the sequel.

Euclid's Second Book

The second book of Euclid's Elements contains an exposition of the basic principles of Geometric Algebra. The first proposition of this book reads:

II,1. If there be two straight lines, and one of them be cut into any number of segments, the rectangle contained by the two straight lines is equal to the rectangles contained by the uncut straight line and each of the segments.

Fig. 49. Diagram to Euclid's Prop. II,1

In modern notation, the theorem is equivalent to the identity

(7)
$$a(b+c+...)=ab+ac+....$$

Propositions II,2 and II,3 are just special cases of II,1. In II,2 the total rectangle is a square, and in II,3 the first partial rectangle ab is a square.

Proposition II,4 is the geometrical equivalent of

(8)
$$(a+b)^2=a^2+b^2+2ab$$

(see Fig. 44). As we have seen, Al-Khwārizmī proves this identity just as Euclid proved it, and he makes use of it in his solution of quadratic equations.

The next two propositions read:

II,5. If a straight line be cut into equal and unequal segments, the rectangle contained by the unequal segments of the whole together with the square on the straight line between the points of section is equal to the square on the half.

II,6. If a straight line be bisected and a straight line be added to it in a straight line, the rectangle contained by the whole with the added straight line and the added straight line together with the square on the half is equal to the square on the straight line made up of the half and the added straight line.

The diagrams illustrating these propositions look very much alike:

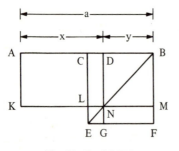

Fig. 50. Euclid II,5

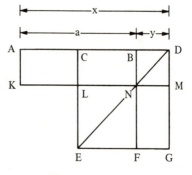

Fig. 51. Euclid II,6

If x and y are the two unequal segments in II,5 ($x>y$), Theorem II,5 can be transcribed by the identity

(9) $$xy+((x-y)/2)^2=((x+y)/2)^2$$

On the other hand, if in Fig. 51 we put $AD=x$ and $BD=y$, we are led to exactly the same identity. This shows that there is something wrong with our transcription. The identity (9) is not an adequate rendering of what Euclid had in mind, for he formulated two theorems and not just one.

In order to obtain a more faithful transcription, let us denote the line segment AB by a. In II.5 as well as in II.6, this segment is diveded into two equal parts $\frac{1}{2}a$ by the point C. In II.5 it is also divided into two unequal parts x and y by the point D, so that

$$x+y=a.$$

On the other hand, in II.6 the segment $BD=y$ is added to a, and the sum AD is our x, so we have in this case

$$x-y=a.$$

If the segment CD is called z in both cases, we have in the first case

$$x=\tfrac{1}{2}a+z$$
$$y=\tfrac{1}{2}a-z,$$

and the theorem II.5 says

(10) $$xy+z^2=(\tfrac{1}{2}a)^2.$$

In the second case we have

$$x=z+\tfrac{1}{2}a$$
$$y=z-\tfrac{1}{2}a,$$

and II.6 says

(11) $$xy+(\tfrac{1}{2}a)^2=z^2.$$

In both cases, a rectangle xy is equated to a difference of two squares: a "gnomon" as the Greeks say (see Fig. 52).

The proof is the same in both cases: from the rectangle xy a smaller rectangle is cut off, and an equal rectangle is added so as to obtain the figure of a gnomon, just as in our Fig. 47.

As we have seen in Chapter 2, the same method of transforming a rectangle into a difference of two squares was also used in the Śulvasūtras. I suppose that this geometrical construction was already used in Pre-Babylonian mathematics.

Now why does Euclid formulate two different theorems, both expressing essentially the same relation between a rectangle and two squares? The answer can be found by asking for what purpose the two theorems are used in the Elements. We shall see that II,5 is used in order to construct two line segments x and y, when the sum $x+y=a$ and the area $xy=C$ are given, whereas II,6 is used when $x-y=a$ and $xy=C$ are given. To see this, we have to discuss the so-called "application of areas with defect or excess".

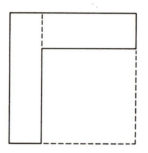

Fig. 52. Gnomon

The Application of Areas

In his Commentary to Euclid I, Prop. 44, Proklos informs us:

These things, says Eudemos (οἱ περὶ τὸν Εὔδημον), are ancient and are discoveries of the Muse of the Pythagoreans, I mean the *application of areas* (παραβολὴ τῶν χωρίων), their *exceeding* (ὑπερβολή) and their *falling-short* (ἔλλειψισ). It was from the Pythagoreans that later geometers took the names, which they again transferred to the so-called *conic* lines, designating one of these a *parabola* (application), another a *hyperbola* (exceeding) and another an *ellipse* (falling-short), whereas these godlike men of old [the Pythagoreans] saw the things signified by these names in the construction, in the plane, of areas upon a finite straight line. For when you have a straight line set out and lay the given area exactly alongside the whole of the straight line, then they say that you *apply* (παραβάλλειν) the said area; when however you make the length of the area greater than the straight line itself, it is said to *exceed* (ὑπερβάλλειν), and when you make it less, in which case, after the area has been drawn, there is some part of the straight line extending beyond it, it is said to *fall-short* (ἐλλείπειν). Euclid too, in his sixth book, speaks in this way both of *exceeding* and *falling short* ... (Translation by Th. Heath: The Thirteen Books of Euclid's Elements I, p. 343).

Proklos is quite right: in Book 6 of the Elements, Propositions 28 and 29, Euclid uses the same terms *application, falling-short,* and *exceeding:*

VI,28. To a given straight line to apply a parallelogram equal to a given rectilinear figure and deficient by a parallelogram similar to a given one.

VI,29. To a given straight line to apply a parallelogram equal to a given rectilinear figure and exceeding by a parallelogram similar to a given one.

The diagrams to these two problems (see Figs. 53 and 54) are quite similar to our diagrams 50 and 51 illustrating II,5 and II,6. In both diagrams, the given line segment is called AB. In VI,28, the problem is, to construct a parallelogram AQ equal to a polygon C, and *deficient* by a parallelogram

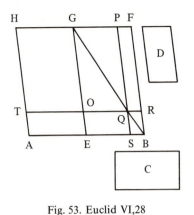

Fig. 53. Euclid VI,28

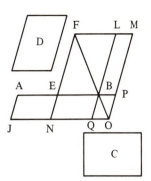

Fig. 54. Euclid VI,29

BQ similar to a given parallelogram D. In VI,29 the parallelogram AO is required to be equal to the polygon C and in *excess* by a parallelogram BO similar to D.

In his paper "Die ersten vier Bücher der Elemente Euklids", Archive for History of Exact Sciences 9, p. 325–380. E. Neuenschwander has shown that the term "parallelogram" was not used in ancient Greek geometry before the time of Eudoxos (about 370 B.C.). In Book 2 of Euclid's Elements only squares and rectangles occur, and the notions "Proportion" and "Similarity" do not occur. So, when the Pythagoreans invented their application of areas with defect or excess, the defect of excess was probably required to be just a square, not a parallelogram similar to a given one. Now if Euclid's diagrams to VI,28 and VI,29 are simplified by assuming a square excess or defect, the resulting diagrams are essentially the same as Euclid's diagrams to II,5 and II,6. Also, the single steps in the proofs of VI,28 and VI,29 are just generalizations of the single steps in the proofs of II,5 and II,6. Thus, one sees that II,5 and II,6 are just the theorems the Pythagoreans needed for the solution of their problems of application of areas with square defect or excess.

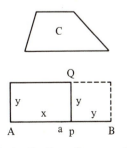

Fig. 55. Application of an area C to a line segment AB with square defect

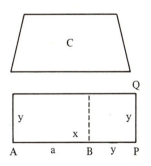

Fig. 56. Application of an area C to a line segment AB with square excess

Let me explain this in greater detail. The application of a given area C to a line segment AB with a square defect may be illustrated by Fig. 55, and with a square excess by Fig. 56. In both cases the rectangle xy is required to be equal to a given polygon C, and the defect or excess, the rectangle BQ is required to be a square. Thus, if the given line segment AB is called a, one has to solve in the case of Fig. 55 the set of equations

(12)
$$xy = C$$
$$x + y = a$$

and in the case of Fig. 56 the set

(13)
$$xy = C$$
$$x - y = a.$$

To solve (12), the Greeks always applied II,5. In Fig. 55 we have

(14)
$$x = \tfrac{1}{2}a + z$$
$$y = \tfrac{1}{2}a - z$$

and II,5 says, as we have seen

(15)
$$xy + z^2 = (\tfrac{1}{2}a)^2.$$

So if $xy = C$ is known, we can solve (15) for z^2:

$$z^2 = (\tfrac{1}{2}a)^2 - C$$

and construct z as a side of a square with given area. Similarly, we have in Fig. 56

(16)
$$x = z + \tfrac{1}{2}a$$
$$y = z - \tfrac{1}{2}a$$

and II,6 says

(17)
$$xy + (\tfrac{1}{2}a)^2 = z^2,$$

so z is again the side of a square with given area.

As we have seen, the problems (12) and (13) are standard problems of Babylonian algebra. To solve (12), the Babylonians make the substitution (14), and to solve (13), they use the substitution (16). In both cases they obtain a pure quadratic equation for z, which can be solved by extracting a square root. Thus we see that the Greek method is essentially the same as the Babylonian method.

In his work "Data", Propositions 84 and 85, Euclid proves: If the rectangle xy is known in mangitude and if $x+y$ or $x-y$ is known, then the line segments x and y are known. The proof is given by an explicit construction of the line segments x and y based on II,5 and II,6. Thus it is clear that Euclid himself uses II,5 to justify the construction of x and y whenever xy and $x+y$ are given, and II,6 whenever xy and $x-y$ are given.

Three Types of Quadratic Equations

The set of equations (12) can be reduced to a single equation by eliminating y, thus obtaining

$$x(a-x)=C$$

or

(18) $$x^2+C=ax.$$

Just so, eliminating y from (13), one obtains

$$x(x-a)=C$$

or

(19) $$x^2=ax+C.$$

One can also eliminate x from (13), thus obtaining

(20) $$y^2+ay=C.$$

Now (18), (19), (20) are just the three types to which all mixed quadratic equations are reduced in the "Algebra" of Al-Khwārizmī. His solutions are equivalent to Euclid's geometrical solutions.

Thābit ben Qurra, who lived in Bagdad in the second half of the 9th century, some 50 years after Al-Khwārizmī, also noted this equivalence. In a very remarkable treatise entitled "On the Verification of the Problems of Algebra by Geometrical Proofs", Thābit shows that the three types of quadratic equations – our types (18), (19), (20) – can be solved by means of the Theorems II,5 and II,6 of Euclid's Elements, and that the geometrical solutions thus obtained agree with the solutions given by "the algebra people" (ahl al-jabr). For each of the three types he compares the single steps of the geometrical solution with the corresponding steps "in the domain of calculation and number", and he shows that there is complete agreement between "what we (geometers) do" and "what they (the algebra people) do". Thābit's little treatise was published and translated into German by P. Luckey in Berichte über die Verhandlungen der sächsischen Akademie der Wissenschaften zu Leipzig 93, p. 93–114 (1941).

Now let's return to Euclid. A typical example of his geometrical solution of quadratic equations is his Proposition II,11. He formulates the
problem to be solved thus:

To divide a given line segment into two parts such that the rectangle contained by the
whole segment and one of the parts is equal to the square on the other part.

If the whole segment is called a and the latter part x, the problem is, to
solve the equation

(21) $$a(a-x)=x^2.$$

Al-Khwārizmī would reduce this equation to its normal form by *al-jabr.*
Adding ax to both sides, he would obtain

(22) $$x^2+ax=a^2.$$

This is an equation of type (20), so it can be solved by applying II,6,
which is in fact what Euclid does. Adding $(\frac{1}{2}a)^2$ to both sides, one obtains

$$(x+\tfrac{1}{2}a)^2=a^2+(\tfrac{1}{2}a)^2.$$

To construct $x+\frac{1}{2}a$, Euclid uses the Theorem of Pythagoras (see
Fig. 57). In a right-angled triangle with sides $\frac{1}{2}a$ and a, the hypotenuse is

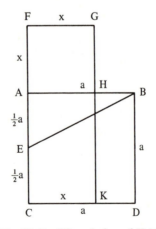

Fig. 57. Euclid's solution of II,11

just $x+\frac{1}{2}a$. Subtracting $\frac{1}{2}a$, Euclid obtains $AF=x$. By applying II,6, Euclid proves that the rectangle FK is equal to the square AB: this is just
what (22) says. Subtracting the rectangle AK from both sides, one obtains
the desired equality (21).

Another Concordance Between the Babylonians and Euclid

In Chapter 2 we have discussed the Problems 8 and 9 of the Old-Babylonian text BM 13901. In Problem 8, the set of equations

(23)
$$x^2 + y^2 = S$$
$$x + y = a$$

is proposed, and in Problem 9 the set

(24)
$$x^2 + y^2 = S$$
$$x - y = a$$

(see my Science Awakening I, p. 70).

Both problems are solved by the "sum and difference method":

Problem 8	Problem 9
$x = \frac{1}{2} a + z$	$x = z + \frac{1}{2} a$
$y = \frac{1}{2} a - z$	$y = z - \frac{1}{2} a.$

In both cases, the auxiliary quantity z is determined from the pure quadratic equation

(25)
$$z^2 + (\tfrac{1}{2} a)^2 = \tfrac{1}{2} S.$$

Now let us compare this solution with the Propositions II,9 and II,10 in Euclid's Elements. The wording of these propositions is similar to that of II,5 and II,6:

II,9. If a straight line be cut into equal and unequal segments, the squares on the unequal segments of the whole are double of the square on the half and of the square on the straight line between the points of section.

II,10. If a straight line be bisected, and a straight line be added to it in a straight line, the square on the whole with the added straight line and the square on the added straight line both together are double of the square on the half and of the square described on the straight line made up of the half and the added straight line as on one straight line.

It would be possible to transcribe both theorems by one and the same modern formula:

(26)
$$x^2 + y^2 = 2 \left\{ \left(\frac{x+y}{2} \right)^2 + \left(\frac{x-y}{2} \right)^2 \right\},$$

but it is better to stick more closely to the geometrical wording and to write, as in the case of II,5 and II,6:

Prop. II,9	Prop. II,10
$x = \frac{1}{2} a + z$	$x = z + \frac{1}{2} a$
$y = \frac{1}{2} a - z$	$y = z - \frac{1}{2} a.$

In both cases, a is the line segment AB in Euclid's diagrams, and z is CD, so x is AD and y is BD (see Figs. 58 and 59).

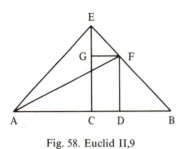

Fig. 58. Euclid II,9

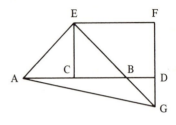

Fig. 59. Euclid II,10

The Theorems II,9 and II,10 can now be transcribed by one and the same formula

$$(27) \qquad x^2 + y^2 = 2\{(\tfrac{1}{2}a)^2 + z^2\}.$$

If we divide both sides by 2, we just obtain the Babylonian formula (25). So the Theorems II,9 and II,10 can be used to justify the Babylonian solutions of the Problems 8 and 9 in BM 13901.

An Application of II,10 to Sides and Diagonals

The Proposition II,10 can also be transcribed thus:

$$(a+y)^2 + y^2 = 2\{(\tfrac{1}{2}a)^2 + (\tfrac{1}{2}a+y)^2\}$$

or more simply, if $\tfrac{1}{2}a$ is called s:

$$(28) \qquad (2s+y)^2 + y^2 = 2s^2 + 2(s+y)^2.$$

Proklos, in his commentary to Plato's "Republic" (ed. Kroll, Vol. 2, § 23) after having explained the Pythagorean theory of "Side-and-Diagonal-Numbers"[6], informs us that the Pythagoreans used the Theorem II,10 in order to prove an "elegant proposition", namely: "If s and y are the side and diagonal of a square, then $s+y$ and $2s+y$ are the side and diagonal of another square."

It is indeed easy to prove this theorem by means of the identity (28). If s and y are side and diagonal of a square, the square on y is twice the square

---

6 The theory of "Side-and-Diagonal-Numbers" will be explained in Chapter 4.

on s (see Plato's dialogue Menon), so the terms y^2 and $2s^2$ in (28) can be cancelled, and we obtain

$$(2s+y)^2 = 2(s+y)^2,$$

hence $s+y$ and $2s+y$ are again side and diagonal of a square.

From the testimony of Proklos we may conclude that the Pythagorean mathematicians knew the Theorem II,10. From Eudemos we know that they also knew the "Application of Areas", which implies that they knew II,5-6. Since the wording of II,9-10 is quite similar to the wording of II,5-6, we may safely conclude that the whole set of four Propositions II,5-6 and II,9-10 was known to the Pythagoreans. Now Book 2, in which the principles of "Geometric Algebra" are systematically developed, is a logical unity, so we may conclude that the whole book is essentially due to the Pythagoreans. For more arguments in favour of this conclusion see Chapter 15 of my book "Die Pythagoreer" (Artemis-Verlag, Zürich 1979).

Thales and Pythagoras

In his commentary to Euclid's Elements, the Neo-Platonist Proklos gives a rapid survey of the history of geometry from Thales to Euclid. The main part of this survey is supposed to come from the "History of Geometry" of Eudemos, written in the school of Aristotle about 300 B.C. This "Catalogue of Geometers" begins thus:

Thales travelled to Egypt and brought geometry to Hellas. He made many discoveries himself, and in many other things he showed his successors the road to the principles.

Several other ancient authors also confirm that the science of geometry originated in Egypt. Herodotos writes:

When the Nile had flooded an agricultural tract, it became necessary for the purpose of taxation, to determine how much land had been lost. From this, to my thinking, the Greeks learnt the art of measuring land. (Herodotos, Histories II 109)

Aristotle says in his "Metaphysics" A 1:

Thus the mathematical sciences originated in the neighbourhood of Egypt, because there the priestly class was allowed leisure.

And Democritos writes:

No one surpasses me in constructing lines with proofs, not even the so-called rope-stretchers (harpedonaptai) of the Egyptians.

Modern authors, including myself, have sometimes adopted a sceptical attitude towards the idea that the mathematical sciences were transmitted from Egypt to Greece. One reason for this scepticism was the fact that the mathematical sciences were on a much higher level in Babylonia than in Egypt. Another reason was the observation that the extant Egyptian texts from the time of the Middle Kingdom contain practical rules for calculating areas and volumes but no proofs and no geometrical constructions.

At present I believe that this scepticism is not quite justified, and that there may be a considerable amount of truth in the statements of Herodotos, Democritos, and Aristotle. First, we must clearly distinguish between a *written* tradition of problems and solutions, such as we find in Babylonian texts and Middle Kingdom papyri, and an *oral* tradition of constructions by means of stretched ropes. As we have seen, the Śulvasūtras do contain proofs for the calculation of areas, and stretched ropes were used in Hindu altar constructions. The Śulvasūtras are written texts, but they are based on a very early oral tradition. It is quite possible that the Egyptian rope-stretchers had similar oral traditions.

The Hindu ritualists, who wrote the Śulvasūtras, were priests, and Aristotle also mentions priests who had leisure to practise geometry. On the other hand, we need not adopt Aristotle's opinion that the mathematical sciences *originated* in Egypt. It seems much more probable that they originated in Neolithic Europe, and they were subsequently transmitted to China, India, Babylonia, Egypt, and Greece.

It is said that Pythagoras travelled from Samos to Miletus to visit Thales, and that Thales advised him to sail to Egypt and to learn from the priests at Memphis and Diospolis (Iamblichos, De vita pythagorica, Greek text and German translation by M. Albrecht, p. 23–25). In Egypt he learnt astronomy and geometry, and was initiated into the mysteries of the gods. Next he came to Babylon, and he was instructed in theology, arithmetic, and musical theory.

In Chapter 1 of my book "Die Pythagoreer" I have tried to judge the reliability of these testimonies. The chronology of the stories recorded by Iamblichos is very doubtful: the duration of his sojourn in Egypt and Babylon is certainly exaggerated. However, it seems quite possible that Pythagoras learnt geometry from Egyptian priests and arithmetic from the Babylonian magi. In Egypt, he may have had contact with the "rope-stretchers", who were experts in geometrical constructions and proofs, and in Babylon he may have acquainted himself with Babylonian arithmetic (including the calculation of Pythagorean triples) and algebra. This would explain the many points of contact between Babylonian algebra and Greek geometric algebra.

The Geometrization of Algebra

Assuming that Pythagoras learnt the solution of quadratic equations and sets of linear and quadratic equations from the Babylonians, and that he transmitted his knowledge to his followers, the question arises: Why did the Greek transform these algebraic methods into the geometric form we find in Euclid's Elements?

One reason for this "geometrization" seems to be the discovery of irrational lines by the Pythagoreans. If the side of a square is taken as a unit of length, the length of the diagonal would have to satisfy the equation

(29) $$x^2 = 2,$$

but this equation cannot be solved in the realm of rational numbers m/n. Approximate solutions of (29) were known to the Greeks as well as to the Babylonians, but the Greeks wanted to have exact solutions. Now in the domain of geometry the equation

$$(30) \qquad\qquad x^2 = C,$$

where C is a given polygonal area, is always solvable: one can construct a square equal to the given polygon. Therefore the Greeks had to transform Babylonian algebra into their geometric algebra.

This explanation, which is due to O. Neugebauer, certainly contains a part of the truth, but it is not the whole truth. We have seen that in the Śulvasūtras a geometrical solution of the equation (30) is given for the case of a rectangle $C = a\,b,$ and that this solution is identical with Euclid's solution. We have also seen that the Śulvasūtras and the Chinese "Nine Chapters" contain geometrical proofs of the same kind as the proofs in Euclid's Book 2. The Hindu ritualists as well as Egyptian rope-stretchers performed constructions by means of stretched ropes, and so did, most probably, the architects of the megalithic constructions in Great Britain. So it seems that a considerable part of the geometry contained in Books 1 and 2 of Euclid's Elements is based on a very early geometrical tradition. Today we have no reason to doubt that Thales and Pythagoras really brought the science of geometry from Egypt to Greece.

Summarizing, we may conclude that the Greeks combined two traditions, which both originated in the Neolithic Age: one tradition of teaching mathematics by means of problems with numerical solutions, and one of geometrical constructions and proofs. The algebraic tradition was mainly transmitted by the Babylonians, and the geometrical tradition probably reached Greece by way of Egypt. Also in the realm of Indo-European languages, a tradition of geometrical altar constructions existed, part of which may well have been preserved by the Greeks in Mycenae and Crete. Traces of this tradition were preserved in Greek stories about the "doubling of the cube". Remember that the origin of this problem was ascribed to King Minos of Crete, who wanted to double the size of an altar without destroying its beautiful form. So we have three possible roads of transmission, namely: by way of Babylon, of Egypt, and of Mycenae and Crete.

"Whatever the Greeks took over from the Barbarians, they made it more beautiful and brought it to perfection", says the author of the Platonic dialogue "Epinomis", the sequel to the "Laws" (Epinomis 988 d–e). He is right.

The Theory of Proportions

Side by side with geometric algebra, the Greeks also had an algebra of proportions. This theory is explained with full proofs in Book 5 of Euclid's Elements. A scholion to this book states that "it is said that the general theory of proportions, explained in this book, was found by Eudoxos".

In Book 5, the theory of proportions is developed not only for geometrical magnitudes, but quite generally for all kinds of magnitudes that have a ratio to one another. Definition 4 explains what this means:

Magnitudes are said to have a ratio to one another which are capable, when multiplied, of exceeding one another.

Thus, two line segments a and b always have a ratio, because a multiple $n \cdot a$ can be made to exceed b and conversely. Just so, two plane areas or volumes have a ratio to each other. Euclid does not say this, but he uses it all the time.

The definition of proportion is as follows:

Definition 5. Magnitudes are said to be in the same ratio, the first to the second and the third to the fourth, when, if any equimultiples whatever be taken of the first and the third, and any equimultiples whatever of the second and the fourth, the former equimultiples alike exceed, are alike equal, or alike fall short of the latter equimultiples respectively, taking in corresponding order.

Thus, the proportionality

$$a:b=c:d$$

means that

$$na>mb \quad \text{implies} \quad nc>md,$$
$$\text{and} \quad na=mb \quad \text{implies} \quad nc=md,$$
$$\text{and} \quad na<mb \quad \text{implies} \quad nc<md.$$

Starting with this definition, Euclid proves a sequence of propositions, culminating in two algebraic rules:

V,12. In a continued proportion

$$a:b=c:d=e:f\ldots$$

the sum of any number of antecedents has the same ratio to the sum of their consequents as any antecedent has to its consequent, e. g.

$$(a+c+e):(b+d+f)=a:b.$$

V,16. If four magnitudes be proportional, they will also be proportional alternately. This means:

$$a:b=c:d \quad \text{implies} \quad a:c=b:d.$$

The whole derivation is a masterpiece of logic.

In earlier times, before Eudoxos, the ratio of two magnitudes a and b was defined by means of the Euclidean algoritm. If a is the larger one of the two segments, one subtracts b from a as often as possible, say n_1 times, until the remainder r is less than b. Now r is subtracted from b as often as possible, say n_2 times, and so on. This process was called *Antanairesis,* which means reciprocal subtraction. The finite or infinite sequence $[n_1,n_2,\ldots]$ was used to define the "Logos" or ratio of the magnitudes a and *b.* The fact that this definition was still in use at the time of Aristotle is proved by a passage from his "Topica" (158b), which reads:

The difficulty in using a figure is sometimes due to a defect in definition, e. g. in proving that the line which cuts the area parallel to one side (of a rectangle or parallelogram) divides similarly the line which it cuts and the area, whereas if the definition be given the fact asserted becomes immediately clear, for the areas have the same antanairesis as have the sides, and this is the definition of "the same ratio".

Fig. 60, explaining a passage from Aristotle's Topica

An early theory of proportions, based upon this definition of ratio, has been reconstructed by O. Becker in his "Eudoxos-Studien" I, Quellen und Studien Gesch. Math. B 2, p. 311–333.

In Book 6, Euclid proves a proposition (VI,14) on parallelograms of equal areas. For the special case of rectangles the proposition says:

If the rectangle spanned by the sides *b* and *c* is equal to the rectangle spanned by *a* and *d,* we have the proportionality

$$a:b=c:d$$

and conversely.

This proposition bridges the gap between Geometric Algebra and the algebra of proportions.

Geometric Algebra in the "Konika" of Apollonios

Apollonios of Perge, the author of the famous "Konika" was a virtuoso on the instrument "Geometric Algebra". In his demonstrations he adds and subtracts and compares not only rectangles and squares, but also parallelograms and triangles, and he makes a free use of proportions.

Let me introduce a convenient notation for proportions. Let a proportion

$$a:b=c:d$$

be given, and let the ratio *a*:*b* be denoted by α. Then we may write

$$c=(a:b)d \quad \text{or} \quad c=\alpha d.$$

In the "Konika", Apollonios deals with three types of conic sections: parabola, hyperbola, and ellipse. For each of the three types he derives a "symptoma": a condition the points of the conic have to satisfy. He chooses a fixed point A on the conic and draws a diameter AA_1 through A (or in the case of the parabola a parallel AA_1 to the axis). Through any point P of the conic he draws a parallel to the tangent at A. This tangent intersects the diameter AA_1 at Q. The line segments $AQ=x$ and $QP=y$ are what we call the "coordinates" of P, in Latin they are called "abscissa" and "ordinate". See Figs. 61 and 62.

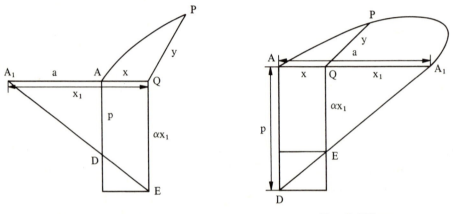

Fig. 61. Hyperbola   Fig. 62. Ellipse

In each of the three cases Apollonios derives a "symptoma": an algebraic relation between x and y. For the parabola the symptoma is quite simple:

$$y^2 = px,$$

where p is a fixed line segment, or in words:
"The square on y is equal to the rectangle spanned by p and x."
For the hyperbola and ellipse the symptomata are more complicated. If $AA_1=a$ is the diameter and if A_1Q is called x_1, we have in the case of the hyperbola (Fig. 61)

(31) $x_1 = x + a$

and in the case of the ellipse (Fig. 62)

(32) $x_1 = a - x.$

In the lower part of the drawing, a fixed line segment $p = AD$ is drawn perpendicular to the diameter AA_1. Let the ratio of p to a be called α. From

Q another perpendicular QE is drawn, which meets the line $A_1 D$ at E. Clearly QE has the same ratio to x_1 as p has to a, so we can write

$$QE = \alpha x_1.$$

The symptoma of the conic section can now be written, in both cases, as

(33) $$y^2 = \alpha x x_1$$

which means

$$y^2 : x x_1 = p : a.$$

If we substitute for x_1 the expression (31) or (32), we obtain for the hyperbola the symptoma

(34) $$y^2 = \alpha x (x + a)$$

and for the ellipse

(35) $$y^2 = \alpha x (a - x),$$

whereas the symptoma of the parabola is, as we have seen,

(36) $$y^2 = p x.$$

If y is given, then $x(x+a)$ is known in the case of the hyperbola, and $x(a-x)$ in the case of the ellipse, and px in the case of a parabola. The determination of x from an equation

$$x(x+a) = C \quad \text{or} \quad x(a-x) = C$$

is what the Greeks call an application of an area with excess or defect, whereas the determination of x from $px = C$ is just an application of a rectangle to a given line segment p without excess or defect.

The names of the three types of conic sections: "parabola", "hyperbola", and "ellipse", are derived from the Greek words for "application", "excess", and "defect".

Apollonios derives the symptoms directly from the definition of the three curves as sections of a cone. For the further development of the theory see O. Neugebauer: Studien zur antiken Algebra II, Quellen und Studien Gesch. Math. B 2, p. 215–254. In this paper, Neugebauer shows clearly how important "Geometric Algebra" is in the work of Apollonios.

The Sum of a Geometrical Progression

In the Babylonian text AO 6484 (see Neugebauer, Mathematische Keilschrifttexte I, p. 99–102) we find the summation of a geometrical progression with ratio 2:

$$1+2+4+\ldots+2^9=2^9+(2^9-1).$$

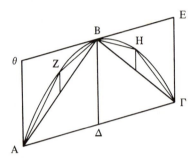

Fig. 63. Quadrature of a parabola segment according to Archimedes

The summation of more general geometrical progressions was known to Archimedes. In his treatise "Quadrature of the Parabola" he determines the area of a parabolic segment $A B \Gamma$ (see Fig. 63) as follows. He first constructs a triangle $A B \Gamma$ by drawing through the midpoint Δ of the base $A \Gamma$ a line ΔB parallel to the axis. Next he constructs in each of the two parabolic segments AZB and $BH\Gamma$, triangles AZB and $BH\Gamma$ in the same manner, and he proves that the sum of these two triangles is equal to $\frac{1}{4}$ of the triangle $A B \Gamma$. If this process is continued, the next step will lead to four triangles, which are together equal to $\frac{1}{4}$ of the sum of the two triangles just considered. The process is continued until what remains of the parabola segment is less than an arbitrarily given area. Obviously, the total area of the triangles that have been removed, is the sum of a finite geometrical progression with ratio $\frac{1}{4}$.

Let A,B,Γ,Δ,E be the terms of this progression, says Archimedes. He now claims that the sum of these terms, increased by $\frac{1}{3}$ of the last term, is exactly $\frac{4}{3}$ of the first term. His proof is very simple:

$$B+B/3 = 4B/3 = A/3$$
$$\Gamma+\Gamma/3 = 4\Gamma/3 = B/3$$
$$\Delta +\Delta/3 = 4\Delta/3 = \Gamma/3$$
$$E+E/3 = 4E/3 = \Delta/3$$

Addition yields

$$B+\Gamma+\Delta+E+B/3+\Gamma/3+\Delta/3+E/3=(A+B+\Gamma+\Delta)/3$$

Adding A and subtracting $B/3 + \Gamma/3 + \Delta/3$ from both sides, one obtains the desired result:

$$(37) \qquad (A + B + \Gamma + \Delta + E) + E/3 = 4A/3 .$$

By a classical "passage to the limit" or "exhaustion argument", Archimedes now concludes that the parabola segment is equal to $4A/3$.

The derivation of Archimedes can be divided into a geometrical and an algebraic part. In the first part, the situation of the triangles with respect to the parabola is important. By means of the "symptoma" of the parabola it is proved that the sum of the triangles AZB and $BH\Gamma$ is just one quarter of the triangle $AB\Gamma$, which implies that the terms $A,B,\Gamma,\ldots$ form a geometrical progression with ratio $1/4$. From now on, the situation of the triangles is unimportant. We have a purely algebraic problem, namely the summation of a geometrical progression.

Just so, several derivations in Apollonios' Konika can be divided into a geometrical and an algebraic part. In the upper half of Fig. 62, the situation of the line segments x and y with respect to the hyperbola matters, but in the lower part of the diagram, only the areas of the rectangles are important. The "symptoma" of the hyperbola says that the square on y is equal to the rectangle AE contained by x and αx_1. The square on y is not even drawn: only its area counts.

Until now, we have discussed some highlights of Greek algebra, namely

Euclid's Book 2 and its application to "Side-and-Diagonal-Numbers",
the Theory of Irrational Lines,
the "Antanairesis", or Euclidean Algorithm, and its application to the definition of ratio,
Eudoxos' Theory of Proportions,
Apollonios' Theory of Conics,
Archimedes' summation of a geometrical sequence.

We now pass to algebraic methods in popular Greek arithmetic.

Sums of Squares and Cubes

In the Babylonian text AO 6484 the sum of the squares of the numbers from 1 to 10 is calculated according to the formula

$$1^2 + 2^2 + \ldots + n^2 = (1 \cdot 1/3 + n \cdot 2/3)(1 + 2 + \ldots + n)$$

for $n = 10$. The Babylonians also knew how to compute the sum of an arithmetical sequence, so they might also have expressed their result as

$$(40) \qquad 1^2 + 2^2 + \ldots + n^2 = (2n + 1)/3 \cdot n(n + 1)/2 .$$

A similar formula was known to the Roman surveyors:

(41) $$1^3 + 2^3 + \ldots + n^3 = \left(n(n+1)/2 \right)^2.$$

Cantor, who records this result in his Geschichte der Mathematik I, p. 559, has no doubt that the Roman surveyors (Agrimensores) had it from a Greek source. In fact, the result (41) is closely related to a set of formulae found in the popular "Arithmetica" of the Neo-Pythagorean Nicomachos of Gerasa:

$$1 = 1^3$$
$$3 + 5 = 2^3$$
$$7 + 9 + 11 = 3^3.$$

If one adds these formulae from 1^3 to n^3, one immediately obtains the formula (41).

Chapter 4

Diophantos and his Predecessors

Part A

The Work of Diophantos

About A. D. 250, an extremely interesting treatise entitled "Arithmeti-ka", dealing with the solution of indeterminate and determinate equations, was written by Diophantos of Alexandria. Originally, it contained 13 books, but in Greek only 6 books are preserved. See the publication of Tannery (Teubner, Leipzig I 1893, II 1895), and the translation and commentary of Th. Heath (second edition, Cambridge 1910, reprinted by Dover, New York 1964).

Quite recently, an Arabic translation of four more books has been discovered. A preliminary translation of these books was presented by J. Sesiano in his Ph D thesis "The Arabic Text of Books IV to VII of Diophantus' Ἀριθμητικά in the translation of Qusta ibn Lūqā" (Xerox University Films, Ann Arbor 1976). A printed edition by the same author entitled "Books IV to VII of Diophantus' Arithmetica" was published by Springer-Verlag, New York 1982.

In what follows the preserved Greek books will be denoted by Greek letters *A,B,Γ,Δ,E,Z,* and the Arabic books by Arabic ciphers 4,5,6,7. According to Sesiano, the original order of the books must have been

$$A \ B \ \Gamma \ 4 \ 5 \ 6 \ 7 \ \Delta \ E \ Z.$$

Diophantos' Algebraic Symbolism

The problems of Diophantos are always stated in words, and most of them involve several unknown numbers. However, by clever artifices, Diophantos manages to reduce them to equations in just one unknown, which he calls ὁ ἀριθμός, "the number". For this unknown number the manuscripts have a special symbol, which in some manuscripts looks somewhat like our letters *s,* and which I shall write as *s* in my transcriptions and comments. Heath has made the suggestion that this symbol is a contraction standing for the first two letters of ἀριθμός. The same symbol is also found in a Michigan papyrus from the second century A. D., which has been published (in translation) by Karpinski and Robbins. The contents of this papyrus will be discussed in Part B of the present chapter.

Diophantos has special names and abbreviations for powers of the unknown s:

$$
\begin{array}{llll}
(1) & \mathring{M} & = \text{Μόνας} & = \text{unity} \\
(s) & \varsigma & = \text{Ἀριθμός} & = \text{number} \\
(s^2) & \varDelta^Y & = \text{Δύναμις} & = \text{square} \\
(s^3) & K^Y & = \text{Κύβος} & = \text{cube} \\
(s^4) & \varDelta^Y\varDelta & = \text{Δυναμοδύναμις} & = \text{square} \times \text{square} \\
(s^5) & \varDelta K^Y & = \text{δυναμόκυβος} & = \text{square} \times \text{cube} \\
(s^6) & K^Y K & = \text{Κυβόκυβος} & = \text{cube} \times \text{cube}
\end{array}
$$

Moreover he has names for reciprocals of powers such as s^{-1}, s^{-2}, etc. For the addition of terms, Diophantos simply writes them in a row, e.g.

$$\varDelta^Y \bar{\gamma} \mathring{M} \overline{\iota\beta} = 3s^2 + 12.$$

Thus he does not simply write 12, but 12 units. In a similar way, Al-Khwārizmī usually writes 12 dirhems.

Diophantos has a definite sign for subtraction, something like an inverted $\mathit{\Psi}$. He says:

A minus (literally a "left over", λεῖψις) multiplied by a minus makes a plus, a minus multiplied by a plus makes a minus, and the sign of a minus is a truncated $\mathit{\Psi}$ turned upside down, thus: $\wedge$.

Diophantos proceeds:

It is well that one who is beginning this study should have acquired practice in addition, subtraction, and multiplication of the various species. He should know how to add positive and negative terms with different coefficients to other terms, themselves either positive or likewise partly positive and partly negative, and how to subtract from a combination of positive and negative terms other terms either positive or likewise partly positive and partly negative.

Next, if a problem leads to an equation in which certain terms are equal to terms of the same species but with different coefficients, it will be necessary to subtract like from like on both sides, until one term is found equal to one term. If by chance there are on either side or on both sides any negative terms, it will be necessary to add the negative terms on both sides, until the terms on both sides are positive, and then again to subtract like from like until one term only is left on each side.

The operations explained here are just the fundamental operations of Al-Khwārizmī: *al-jabr* and *al-muqābala*. *Al-jabr* is the addition of equal terms to both sides of an equation in order to eliminate negative terms. *Al-muqābala* is the subtraction of equal terms from both sides of an equation.

If one has luck, these operations leave only one positive term on each side of an equation. If the highest power of s occurring on either side is s or s^2, the equation then takes the form

$$as=b \quad \text{or} \quad as^2=bs \quad \text{or} \quad as^2=b$$

and it is easy to solve.

Diophantos promises to discuss at a later point the case, in which two terms remain on one side. In the part of his work that has been preserved, this promise is not fulfilled. Still, it becomes evident that he knew how to solve quadratic equations like

$$as^2+bs=c \quad \text{or} \quad as^2+c=bs \quad \text{or} \quad bs+c=as^2$$

as well as the Babylonians, the Indians, and the Arabs.

The 6 forms of equations of the first and second degree, mentioned here, are exactly the 6 standard forms of Al-Khwārizmī.

Determinate and Indeterminate Problems

A single equation of degree n in one unknown cannot have more than n solutions in any field. Just so, a set of k equations in just as many unknowns usually has only a finite number of solutions. Problems leading to equations of this kind are called *determinate*.

On the other hand, a set of k algebraic equations in more than k unknowns very often has an infinite number of rational solutions. If a problem leads to equations of this kind it is called *indeterminate*.

In Babylonian algebra, and in Book A of Diophantos, most problems are determinate. We shall see that the methods by which Diophantos solves such problems are closely related to Babylonian methods. It seems to me that this part of the work of Diophantos is just a continuation and refinement of a very old tradition, which is represented not only by the Babylonian and Chinese texts discussed in Chapter 2, but also by Egyptian and Greek texts of the Hellenistic and Roman period. Without going into details, I shall mention only two of these texts, one Demotic and one Greek:

1. A Demotic papyrus from Hermopolis West, now in the Cairo museum, written in the third century B.C. and published by R.A. Parker[7], contains 40 mathematical problems. The very remarkable problems 34 and 35 deal with rectangular plots of land of which the area and the diagonal are given. Each of these problems leads to a pair of equations of the form

$$x^2+y^2=d^2$$
$$xy=A.$$

---

7 R.A. Parker: Demotic Mathematical Papyri. Brown University Press, Providence R.I., and Lund Humphreys, London 1972.

These problems are of just the same kind as Diophantos' problems 27–30 of Book A, and the method of solution is also similar, with the important difference that Diophantos always motivates his method of solution, whereas the Cairo papyrus only gives rules for calculating the solutions.

2. The Greek papyrus Michigan 620 published by Karpinski and Robbins contains two problems leading to sets of linear equations. It is very remarkable that this papyrus uses the same sign for an unknown number as Diophantos.

The examples just given show clearly that, as far as *determinate* problems are concerned, Diophantos continues an early tradition, which was still living in his own country, in Hellenistic Egypt. On the other hand, his methods of solution of *indeterminate* problems seem to be new, and highly original.

I shall now give some examples of Diophantos' problems and methods of solution.

From Book A

Problems 27–30 of Book A read in the translation of Heath:

27. To find two numbers such that their sum and product are given numbers.

Necessary condition: The square of half the sum must exceed the product by a square number.

28. To find two numbers such that their sum and the sum of their squares are given numbers.

Necessary condition: Double the sum of their squares must exceed the square of their sum by a square.

29. To find two numbers such that their sum and the difference of their squares are given numbers.

30. To find two numbers such that their difference and product are given numbers.

Necessary condition: Four times the product together with the square of the difference must give a square.

Problem 29 can be solved rationally. In the three other cases, a condition is necessary to ensure the existence of a rational solution, and Diophantos states this condition clearly.

The method of solution is just the "Sum and Difference" method we have also found in Babylonian texts (see Chapter 2). Whenever the sum $x+y=a$ is given, Diophantos puts

$$x=\tfrac{1}{2}a+s, \quad y=\tfrac{1}{2}a-s,$$

and when the difference $x-y=d$ is given, he puts

$$x=s+\tfrac{1}{2}d, \quad y=s-\tfrac{1}{2}d.$$

As an example, I shall reproduce the solution of problem 28:

Let the sum of the numbers be 20, and the sum of their squares 208.

Let their difference be $2s$. The larger one will be $10+s$, because half of the sum is 10, and the smaller one $10-s$, for then the sum is 20 and the difference $2s$. It remains to make the sum of their squares equal to 208. But the sum of the squares is $2s^2+200$. This must be equal to 208, hence s becomes 2. Returning to the original hypothesis, we find that the larger number will be 12, the smaller one 8, they fulfill the conditions of the proposition.

Note that the problem is of the same type as problem 8 of the Babylonian text BM 13901:

I have added the areas of my two squares: 21,40. I have added the sides of my squares: 50.

The Babylonian method of solution is the same as in Book A of Diophantos. The given sum 50 is halved: result 25. The sides of the two squares are obtained as $25+s$ and $25-s$. The square of s is calculated as

$$s^2 = \tfrac{1}{2} \cdot 21{,}40 - 25^2$$
$$= 10{,}50 - 10{,}25 = 25 \, ,$$

hence s is 5, and the two sides are

$$25 + 5 = 30 \quad \text{and} \quad 25 - 5 = 20 \, .$$

The problems 27 and 30 are standard problems of Babylonian algebra. See H. Goetsch: Die Algebra der Babylonier, Archive for History of Exact Sciences 5, section 5.2 on p. 119-130. Once again, Diophantos' method of solution is the same as the Babylonian standard method, but Diophantos gives a complete justification of the method.

From Book B

Problem 20. To find two numbers such that the square of either added to the other gives a square.

Let the first number be s, the other $2s+1$, then the square of the first plus the second gives a square. The square of the second plus the first is $4s^2+5s+1$. This must be equal to a square. I form this square from $2s-2$, then it is $4s^2+4-8s$, and s becomes 3/13. The first number is 3/13, the second 19/13.

Most problems in Book B are solved by a similar method. If one has to find two or three numbers satisfying certain conditions, the numbers are assumed to be simple expressions in s satisfying identically all conditions of the problem but one. In our case the expression are s and $2s+1$, and the condition

$$s^2 + (2s+1) = \text{square}$$

is satisfied identically in s. The remaining condition requires that

$$(2s+1)^2 + s = 4s^2 + 5s + 1$$

be a square. If this square is called (in modern notation) t^2, one has to solve an equation of the form

(1) $$as^2+bs+c=t^2$$

by rational numbers s and t.

Diophantos has several methods to deal with an equation of the form (1).

First method. If a is a square, say $a=e^2$, Diophantos puts

(2) $$t=es+m,$$

where m is chosen so as to yield a positive solution. If (2) is substituted into (1), the term as^2 cancels out, and one obtains a linear equation for s.

Second method. If c is a square, say $c=f^2$, the substitution

(3) $$t=f+ms$$

leads to a quadratic equation in s of the form

(4) $$gs^2=hs$$

which can be solved rationally:

$$s=h/g.$$

Third method. This method can be applied if the middle term bs in (1) is missing, and $a+c$ is a square. Diophantos explains his third method in a lemma preceding problem 12 of Book Z. The lemma reads in the translation of Heath:

Lemma 2 to the following problems.

Given two numbers (a and c in our notation) the sum of which is a square, an infinite number of squares (x^2) can be found such that, when the square is multiplied by one of the given numbers and the product is added to the other, the result is a square (i.e. such that ax^2+c is a square).

The problem for which Diophantos wants to construct an infinity of solutions can be formulated in modern notation as

(5) $$ax^2+c=y^2.$$

Diophantos makes the substitution

$$x=s+1 \quad \text{hence} \quad x^2=s^2+2s+1,$$

and he finds an equation

$$as^2+2as+(a+c)=y^2$$

which can be solved by the second method, because $a+c$ is a square.

Diophantos gives the proof of his lemma only for the special case $a=3$, $c=6$, but the idea of the proof is perfectly general. He says:

Let the two given numbers be 3 and 6. We have to find a square which, if we multiply it by 3 and add 6, becomes a square.

Let s^2+2s+1 be the required (first) square, then $3s^2+6s+9$ must be equal to a square. Since the number of units is a square, an infinite number of solutions can be found. For instance, we may put

$$3s^2+6s+9=(3-3s)^2,$$

and s becomes 4. The required square is the square of 5, and an infinite number of other solutions can be found.

Fourth method. This method, a generalization of the third, is explained in a lemma to problem 15 of Book Z. Diophantos considers an equation

(6) $ax^2-c=y^2$

and he says: if one has a particular solution of the problem, say

$$x=d, \quad y=e,$$

one can always find another solution larger than this one. Such a larger solution is found by putting

(7) $x=d+s, \quad y=e+ms.$

One obtains an equation of the form (4) for s, which can be solved rationally.

In modern terminology, (6) is the equation of a hyperbola. If one point (d,e) on the hyperbola is given, one can draw a straight line through this point with arbitrary slope m. A parametric representation of this line is given by (7). The line intersects the hyperbola in a second point, which can be calculated rationally.

What we today call "equation" of the hyperbola is just what the Greek call its "symptoma". Apollonios would have no difficulty in following my geometrical interpretation of the method of Diophantos. I guess, Diophantos would have no difficulty either.

The Method of Double Equality

When Diophantos wants two expressions in s to be squares, he often uses the "Method of Double Equality". This very ingenious method is based on the identity

$$x^2-y^2=(x+y)(x-y).$$

Whenever two expressions in s have to be squares, their difference has to be equal to the difference of two squares (x^2-y^2). This is accomplished by factoring the difference and equating the factors to $x+y$ and $x-y$. Half the sum of these factors is x, and half the difference is y.

The method is applied for the first time in problem 11 of Book B:

Problem 11. To add one and the same number to two given numbers so as to make each of them a square.

Let the given numbers be 2 and 3, and let the added number be s. Therefore $s+2$ and $s+3$ must be squares. This is called a *double equality*. Take their difference and look for two numbers whose product is equal to this, say 4 and 1/4. Then take either the square of half of the difference between the two factors and equate it to the lesser expression, or the square of half the sum and equate it to the greater. In the first case the square of half the difference is 225/64. Therefore

$$s+2=225/64$$

and s becomes 97/64. On the other hand, the square of half the sum is

$$s+3=289/64$$

which gives again $s=97/64$.

From Book Γ

I shall give an example in Diophantos' own wording.

Problem 10. To find three numbers such that the product of any two of them, increased by a given number, produces a square.

Let the given number be 12. The product of the first and second number plus 12 must be a square, say 25. If from this we subtract 12, we obtain the product of the first and second number: 13. Therefore set the first number equal to $13s$, the second to s^{-1}.

Again subtract 12 from another square, say 16, and let the difference ($=4$) be the product of the second and third numbers. But the second number is s^{-1}, hence the third is $4s$.

The product of the first and third plus 12 must also be a square. The product of the first and third is $52s^2$, hence $52s^2+12$ must be a square. If the factor 52 were a square, it would be easy to satisfy this condition. Now 52 was found as the product of 13 and 4; if these were both squares, their product would be a square. Thus we have to find two squares, each of which would produce a square if increased by 12. This is easy; for example, we can choose 4 for one of the squares and 1/4 for the other, since each of them, increased by 12 is a square.

Let us therefore start all over again and set the first number equal to $4s$, the second to s^{-1}, and the third to $(1/4)s$. We still have to make sure that the product of the first and third plus 12 is a square. The product of the first and third is s^2, hence s^2+12 must be a square. Choose as the side of the square $s+3$, then the square becomes s^2+6s+9 and we find $s=1/2$. This solves the problem.

What Diophantos applies here is the "method of wrong hypothesis"; he starts by assuming certain expressions for his unknown numbers, and, when this brings him into trouble, he looks for the causes of the difficulty and for the changes in the assumptions which will enable him to solve the problem.

In the present case $52s^2+12$ would be a square for $s=1$. Diophantos seems to have overlooked this easy solution.

From Book 4

Book 4 is available only in the Arabic translation by Qustā ibn Lūqā of Baalbek. The problems solved in this book are mainly concerned with squares and cubes. Diophantos always tries do reduce the equations to the simple form

$$a s^m = b s^n.$$

If m and n differ by just 1 unit, the equation has a rational solution. In some cases m and n differ by 2 or 3 units; in these cases one must require that a/b be a square or a cube.

I shall give a simple example in the translation of Sesiano, replacing Sesiano's letter x by an s:

Problem 3. We wish to find two square numbers, the sum of which is a cubic number.

We put s^2 as the smaller square and $4s^2$ as the greater square. The sum of the two squares is $5s^2$, and this must be equal to a cubic number. Let us make its side any multiple of s we please, say s again, so that the cube is s^3. Therefore, $5s^2$ is equal to s^3. As the side which contains the s^2 is less in degree, we shall divide the whole by s^2; hence s is equal to 5. Now, since we assumed the smaller square to be s^2, and s^2 arises from the multiplication of s – which we found to be 5 – by itself, s^2 is 25. And, since we put for the greater square $4s^2$, it is 100. The sum of the two squares is 125, which is a cubic number with 5 as its side.

For most problems of Book 4, the method is quite similar. In some cases, involving a set of two conditions, the "method of double equality" is applied.

From Book 5

Most problems of Book 5 involve higher powers of the unknown numbers. The methods of solution are essentially the same as in Book 4, but the calculations are more complicated.

Problems 7–12 give rise to pairs of equations with two unknowns, such as

Problem 7
$$x + y = 20$$
$$x^3 + y^3 = 2240$$

Problem 8
$$x - y = 10$$
$$x^3 - y^3 = 2170$$

Problem 9
$$x + y = 20$$
$$x^3 + y^3 = 140(x - y)^2$$

Problem 10
$$x - y = 10$$
$$x^3 - y^3 = (8 + 1/8)(x + y)^2$$

Problem 11
$$x - y = 4$$
$$x^3 + y^3 = 28(x + y)$$

Problem 12

$$x+y = 8$$
$$x^3 - y^3 = 52(x-y).$$

The method of solution is always the same. If the sum $x+y=a$ is given, one puts $x=\frac{1}{2}a+s$ and $y=\frac{1}{2}a-s$. If $x-y=d$ is given, one sets $x=s+\frac{1}{2}d$ and $y=s-\frac{1}{2}d$. In each case one gets a pure quadratic equation for s, which has a rational solution only if the given numbers satisfy a certain condition.

As an example, I shall here reproduce the wording of problem 8, the wording of the necessary condition, and the beginning of the solution:

Problem 8. We wish to find two numbers such that their difference and the difference of their cubes are equal to two given numbers.

It is necessary that four times the number given for the difference of the cubes exceed the cube of the number given for the difference of the two numbers by a number which, when divided by three times the number given for the difference of the two numbers, gives a square, and which, when multiplied by three quarters of the number belonging to the said difference, gives a square.

Let the number given for the difference of the two numbers be 10 and the number given for the difference of the two cubes be 2 170. We wish to find two numbers having 10 as their difference and 2 170 as the difference of their cubes. We put $2s$ as the sum of the two numbers, so that one will be $s+5$ and the other, $s-5$: this, in order that their difference amount to 10. We form the cube of each of them.

The rest is obvious. Note that the two necessary conditions are equivalent. If $a^2/3b$ is a square, $a^2 \cdot (3/4)b$ is also a square. Possibly the superfluous second condition is due to a scholiast. Quite generally, the Arabic text is more prolix than the Greek text usually is.

From Book 7

Problem 15. We wish to divide a given square number into four parts such that two of the four parts each give when subtracted from the given square number a square, and the other two each give when added to the given square number a square.

Let the given square number be 25. The problem is equivalent to a set of five conditions for four unknown numbers z, t, x, y:

(8)	$x+y+z+t=a^2=25$
(9)	$a^2+x=\text{square}$
(10)	$a^2+y=\text{square}$
(11)	$a^2-z=\text{square}$
(12)	$a^2-t=\text{square}.$

Diophantos first replaces the given square number a^2 by an unknown square s^2. He next puts $x=2s+1$, and $y=4s+4$, so that conditions (9) and (10) are satisfied identically. He also puts $z=2s-1$ and $t=4s-4$, so that conditions (11) and (12) are satisfied identically. Now $x+y$ is $6s+5$, and $z+t$ is $6s-5$, hence

$$s^2=x+y+z+t=12s,$$

so s is 12, and s^2 is 144. The additive parts of 144 are

$$2s+1=25$$
$$4s+4=52$$
$$2s-1=23$$
$$4s-4=44.$$

Now Diophantos says: If the given square number were 144, we would have reached our goal, but is was 25. Consequently, we shall have to multiply each of the parts of 144 by 25 and divide the result by 144. The result is

$$x=25\cdot\frac{25}{144}=\frac{625}{144}$$
$$y=52\cdot\frac{25}{144}=\frac{1\,300}{144}$$
$$z=23\cdot\frac{25}{144}=\frac{575}{144}$$
$$t=44\cdot\frac{25}{144}=\frac{1\,100}{144}.$$

From Book $\varDelta$

A nice example of the "method of the wrong assumption" is the following:

Problem 31. To divide unity into two parts, such that, if given numbers are added to them respectively, the product of the two sums gives a square.

Let 3 and 5 be the added numbers. Let one part be s, the other $1-s$. Addition of 3 and 5 to these parts gives $s+3$ and $6-s$. Their product is therefore $3s+18-s^2$. This has to be equal to a square. Let it be equal, e.g., to $4s^2$, so that one finds

$$3s+18=5s^2.$$

This equation cannot be solved in rational numbers. The coefficient 5 was a square plus 1. In order that the equation may have rational solutions, this coefficient multiplied by 18, increased by the square of one half of 3, must be a square. Thus we have to look for a square which, increased by 1, and multiplied by 18, and then increased by $2+1/4$, produces a square.

(Introduction of a new number s:)

Let s^2 again be the required square. Then s^2+1 times 18 plus $2+1/4$, or $18s^2+20\frac{1}{4}$ must be a square. Take this four times: $72s^2+81$ has to be a square. Set it equal to $(8s+9)^2$; one finds $s=18$. The required square is therefore 324.

Now we return to the original question: $3s+18-s^2$ must become a square. Take for it $324s^2$; then we find $s=78/325=6/25$. The first part becomes 6/25, the second 19/25.

Clearly Diophantos knew how to solve the general quadratic equation of the form

$$as^2 = 2bs + c.$$

He knew that the solution is rational if $b^2 + ac$ is a square.

Diophantos gives a second solution for the same problem, in which he operates with inequalities:

Differently. Let the first part be s less the 3 units which have to be added to it. Then the second part is $4-s$, and the condition becomes: $9s-s^2$ must be a square. Set this square equal to $4s^2$; then one finds $s=9/5$. But we cannot subtract 3 units from this; s has to be made greater than 3, but less than 4.

Now s was found by dividing 9 by 5, e.g. by a square plus 1. If s lies between 3 and 4, then the square plus 1 must lie between 3 and $2+1/4$, and the square between 2 and $1+1/4$. Therefore we have to look for a square that is greater than $1+1/4$ but less than 2.

Reduce these numbers to the denominator 64, so that the numerators become 80 and 128. Now it is easy: the square is 100/64, i.e. 25/16.

Returning to the original problem, we find the equation

$$9s-s^2 = (25/16)s^2,$$

from which follows $s=144/41$. Therefore the first number will be 21/41, the second 20/41.

From Book E

A simple example of the "method of the double equation":

Problem 1. To find three numbers in geometrical progression such that each of them decreased by 12 gives a square.

Look for a square which, decreased by 12, again produces a square.

This is easy: the square is $42+1/4$.

Now let the first term of the progression be $42+1/4$, the third term s^2; then the middle term is $6\frac{1}{2}s$. Now each of the two expressions s^2-12 and $6\frac{1}{2}s-12$ has to be a square. Their difference is $s^2-6\frac{1}{2}s$. Resolve this in two factors: s and $s-6\frac{1}{2}$. Half their difference multiplied by itself is 169/16; we equate this to the smallest expression $6\frac{1}{2}s-12$. Thus we get $s=361/104$, etc.

Book Z deals with Pythagorean triangles. We shall discuss the contents of this book in the next chapter.

Part B

The Michigan Papyrus 620

At the beginning of the present chapter, I have mentioned the fact that the special sign for the unknown s, which looks somewhat like a final sigma in some of the manuscripts of the "Arithmetika", is also found in the Michigan Papyrus 620. An English translation of this papyrus was given by

L. C. Karpinski and F. E. Robbins in Science 70 (1929) p. 311–314. The papyrus was written, most probably, in the second century A. D.

The papyrus presents three arithmetical problems, each leading to a set of linear equations in two, three, or four unknowns. The first problem reads in the reconstruction of Karpinski and Robbins:

Four numbers: their sum is 9900; let the second exceed the first by one seventh of the first; let the third exceed the sum of the first two by 300, and let the fourth exceed the sum of the first three by 300; to find the numbers ...

The text explains the method of solution in words. Only the last part of this explication is preserved, followed by a check of the correctness of the solution. After this check, the whole calculation is summarized as follows:

1/7			300	300	9900
7s	8s		15s+300	30s+600	
1050	1200		2550	5100	
150					

The first line contains the data of the problem. In the second line, the first unknown number is put equal to 7s. The second number is then 8s, the third 15s+300, and the fourth 30s+600. The sum of the four numbers is now

$$60s+900=9900$$

from which one obtains $s=150$. This number is entered in the last line. Substituting this value of s, one obtains the four required numbers

$$1050, \quad 1200, \quad 2550, \quad 5100,$$

which indeed satisfy all conditions.

The other two problems are quite similar. As Karpinski and Robbins rightly remark: "In the form of solution of these algebraic problems we have a remarkable approach to modern algebraic symbolism."

From this papyrus we see that Diophantos' method of reducing algebraic problems to equations with only one unknown quantity s was known already in the second century A. D. In the papyrus, this method is applied to definite problems only, whereas Diophantos applies it to indefinite problems as well.

Part C

Indeterminate Equations in the Heronic Collection

In a Byzantine manuscript (probably of the twelfth century) from which H. Schöne edited the "Metrika" of Heron of Alexandria we find a collection of indeterminate problems with solutions. They have been published and translated by Heiberg in Bibliotheca mathematica 8 (1907) p. 118–134, with comments by Zeuthen. A good summary can be found in Th. Heath: A History of Greek Mathematics II, p. 444–448.

The first problem is, to find two rectangles such that the perimeter of the second is three times that of the first, whereas the area of the first is three times that of the second. Replacing the given number 3 by an arbitrary integer n, we have to solve a set of two equations in four unknowns:

(1) $$u+v=n(x+y)$$

(2) $$xy=n \cdot uv.$$

The solution given in the text is

$$x=2n^2-1$$
$$y=2n^3$$
$$u=n(4n^3-2)$$
$$v=n.$$

Zeuthen suggests that the solution may be explained thus. Any solution may be multiplied by an arbitrary factor, so we may put $v=n$. We may also put $u=nz$, where z is a new unknown. We now have the simpler set

(3) $$x+y=1+z$$

(4) $$xy=n^3z.$$

Eliminating z, one obtains

$$xy=n^3(x+y)-n^3$$

or

$$(x-n^3)(y-n^3)=n^3(n^3-1).$$

An obvious solution is

$$x-n^3=n^3-1$$
$$y-n^3=n^3$$

or

$$x=2n^3-1$$
$$y=2n^3$$
$$z=x+y-1=4n^3-2,$$

etc.

Another possibility, more in the line of Diophantos, would be as follows. If $x+y$ and xy are given as in (3) and (4), it is natural for Diophantos as well as for the Babylonians to calculate $y-x$ by means of the identity

$$(y-x)^2 = (x+y)^2 - 4xy.$$

In our case the calculation yields

$$(y-x)^2 = (1+z)^2 - 4n^3 z$$
$$= z^2 - (4n^3 - 2)z + 1.$$

This quadratic expression in z must be equal to a square. As we have seen, Diophantos has several methods to satisfy conditions of this kind. In our case, the simplext method is, to put the square equal to 1. This leads to the equations

$$z^2 = (4n^3 - 2)z$$
$$y - x = 1$$

and hence to the solution given in the text.

The second problem is similar. It leads to the two equations

(5) $$x + y = u + v$$
(6) $$xy = n \cdot uv.$$

The solution given in the text is

$$x = n^2(n-1)$$
$$y = n^2 - 1$$
$$u = n - 1$$
$$v = n(n^2 - 1).$$

A more general solution can be obtained by the substitution

(7) $$v = my.$$

Substituting this into (6) and dividing by y, one obtains

(8) $$x = mnu.$$

Substituting (7) and (8) into (5), one obtains

$$(mn - 1)u = (m - 1)y.$$

The simplest solution of this equation is

$$u = m - 1$$
$$y = mn - 1$$

from which one obtains

$$v = m(mn - 1)$$
$$x = mnu = mn(m - 1).$$

The special case $m = n$ yields the solution given in the text.

Chapter 5

Diophantine Equations

Diophantine equations in the modern sense are equations to be solved by integers, such as the linear Diophantine equation

(A) $$ax + c = by$$

or "Pell's Equation"

(B) $$x^2 = Dy^2 + 1$$

or the equation of "Pythagorean triples"

(C) $$x^2 + y^2 = z^2.$$

Diophantine equations are called after Diophantos of Alexandria, but Diophantos himself considers mainly equations to be solved by *rational* numbers m/n. In the present chapter, we are concerned only with *integer* solutions of the three types (A), (B), (C). Accordingly, this chapter will be divided into three parts A, B, and C.

In transcribing Sanskrit names and words like Aryabhata and Siddhanta, I shall leave out all bars and dots. This will not give rise to misunderstandings.

Part A

Linear Diophantine Equations

The earliest known systematic treatment of Diophantine equations of the form (A) is found in the treatise "Aryabhatiya" of the famous Indian astronomer Aryabhata[8]. Roger Billard[9] has shown that Aryabhata's astron-

---

8 This treatise has been translated by W. E. Clark: The Aryabhatiya of Aryabhata, Chicago 1930, and by K. S. Shukla: Aryabhatiya of Aryabhata, New Delhi 1976, with a valuable commentary.
9 R. Billard: L'astronomie indienne, Paris 1971.

omical theory was based on observations made about +510. Aryabhata himself informs us that he was just 23 years old when 3600 years of the current Kaliyuga had elapsed, that is in +499. So the date of Aryabhata is completely certain.

The Kaliyuga is the last of the four quarteryugas of which Aryabhata's fundamental return period of 4320000 years is composed. The Kaliyuga was supposed to begin on February 18, 3102 B.C.

The Aryabhatiya consists of four parts. In the first part, the astronomical system of Aryabhata is explained in a very concise form. The second part deals with *Ganita* or mathematics. It contains two stanzas 32 and 33 dealing with the solution of linear Diophantine equations. These stanzas are extremely condensed and difficult to understand. To get their meaning, one has to compare Aryabhata's text with the the explanations of his commentators Bhaskara I (+629) and Parameshvara (+1431) and with the texts of Aryabhata's successors Brahmagupta (+628) and Bhaskara II (+1150). The mathematical parts of the astronomical treatises of Brahmagupta and Bhaskara II are available in the English translation of L.C. Colebrooke: Algebra with Arithmetic and Mensuration from the Sanskrit of Brahmegupta und Bháscara, London 1817, reprinted Wiesbaden 1973 (Martin Sändig). In what follows, all English translations from texts of Brahmagupta and Bhaskara II will be taken from the treatise of Colebrooke.

From the explanations of the commentators we may conclude that Aryabhata deals with a pair of linear equations of the form

(1)
$$a_1 x_1 + c_1 = b_1 y$$
$$a_2 x_2 + c_2 = b_2 y$$

in which x_1, x_2, and y are required to be integers. Parameshvara shows that the solution of such a pair of equations can be reduced to the successive solution of three single equations of type (A). So let me first explain the method of solving an equation (A), as presented by our Sanskrit authors.

In what follows I shall try to reproduce, as far as possible, the original wording of the Hindu authors, in the English translations of Clark, Shukla, and Colebrooke. For the reader's convenience, I shall also use our algebraic notation, but only in those cases in which the modern formulae can be re-translated directly into the original terminology.

If one wants to solve an equation (A), the first step is, to determine the largest common divisor d of the coefficients a and b by means of the Euclidean algorithm. If d does not divide the "additive" c, the problem is "ill put", or impossible. If it does divide c, all terms of the equation can be divided by d. Bhaskara II says: "Being divided by that common measure, they (the coefficients) are termed reduced quantities."

So we may suppose $d = 1$. The next step is to divide b by a with a remainder $r < a$:

(2) $b = qa + r$.

which has only one solution x less than 78. Namely: q is the quotient of the division of 576 by 78, and x is the remainder.

The method of calculation is presented, as usual in the "Nine Chapters", in the form of a general rule.

The problems 38–43 are all of the same kind. The problems 44–46 are similar, but now the prices pro unit are lower than 1 coin, so that one has to ask: How many units can one get for one coin? Mathematically, the problem is just the same: one now has to divide the number of units by the number of coins, so as to obtain a quotient and a remainder.

In all these cases, the solution of the Diophantine equation (27) is very simple, because one of its coefficients, namely $p - q$, is 1. One division is sufficient to yield the desired result.

However, the Chinese were also able to solve less simple Diophantine problems, as we shall see in the next section.

The Chinese Remainder Problem

"Master Sun's Arithmetical Manual" *(Sun Tzu Suan Ching)* is one of the earliest preserved Chinese textbooks on Arithmetic. According to J. Needham (Science and Civilisation in China, Vol. 3, p. 33) the manual was written between $+280$ and $+473$. For a description of its contents I may refer to Y. Mikami: The Development of Mathematics in China and Japan (second ed., Chelsea Publishing Co., New York 1974).

According to Mikami, the subject matter of Sun's manual is mainly the same as that of the "Nine Chapters", but there is one problem that first appears in Sun's manual, namely:

We have a number of things, but do not know exactly how many. If we count them by threes we have two left over. If we count them by fives we have three left over. If we count them by sevens we have two left over. How many things are there?

Sun-Tzu gives not only the solution of this particular problem, with remainders 2,3,2, but also a general rule for arbitrarily given remainders. He instructs the reader first to take a multiple of 5×7 which leaves a remainder 1 when divided by 3, namely 70, next a multiple of 3×7 leaving a remainder 1 when divided by 5, namely 21, and next a multiple of 3×5 leaving a remainder 1 when divided by 7, namely 15. Next he calculates

$$2 \times 70 + 3 \times 21 + 2 \times 15 = 140 + 63 + 30 = 233.$$

This, obviously, is a solution of the problem. The least solution is obtained by casting out a multiple of

$$3 \times 5 \times 7 = 105$$

as often as possible. Result 23.

It is very remarkable that exactly the same problem also occurs as Problem 5 of the "supplementary problems" printed by Hoche in his edition of the "Introduction to Arithmetic" of Nikomachos of Gerasa. According to Needham (Science and Civilisation, Vol. 3, p. 34, footnote a) this problem occurs in only two or three of the nearly fifty extant manuscripts of Nikomachos. Three of the five "supplementary problems" are ascribed to the Byzantine monk Isaac Argyros, who flourished in the 14th century. Therefore it is reasonable to assume that Problem 5 was added in the 14th century in the school of Isaac Argyros.

It is very improbable that the Byzantine scribe was influenced by Chinese mathematics. In all probability we have to assume a common origin for the Greek and the Chinese text: either an ancient Greek text or a still more ancient pre-Greek tradition.

We have seen that Brahmagupta, in Chapter 18 of his Siddhanta, discusses the problem: "What number, divided by 6, has a remnant of 5, and by 5, a remnant of 4, and by 4, one of 3, and by 3, one of 2?" Bhaskara II shows that this problem can be solved by means of the "pulverizer". See Colebrooke: Algebra with Arithmetic and Mensuration, p. 235–236.

The question now arises: Why were Sun-Tzu, Aryabhata, Brahmagupta, Isaac Argyros, and Bhaskara II interested in the solution of problems of this kind?

Astronomical Applications of the Pulverizer

Aryabhata was primarily an astronomer, and so were Brahmagupta and Bhaskara II. In the Aryabhatiya, the section on the pulverizer (Ganitapada 32–33) is preceded and followed by sections concerning astronomical problems. Brahmagupta gives, in his Chapter 18, several applications of the pulverizer to astronomical problems: they are concerned with numbers or revolutions of planets in a *yuga*. i. e. in a large astronomical period. Therefore we may conjecture that Brahmagupta was interested in the pulverizer just because of its importance in astronomy.

This conjecture will be confirmed by an analysis of the astronomical systems of Aryabhata and Brahmagupta.

Aryabhata's Two Systems

In his very interesting paper "Aryabhata, the Father of Indian Epicyclic Astronomy" (J. of the Department of Letters, University of Calcutta, Vol. 18, p. 21–56, 1929), P. C. Sengupta has shown that Aryabhata developed two astronomical systems, differing only in details: the *Midnight System* and the *Sunrise System*. The latter system is known from Aryabhata's own treatise Aryabhatiya, the former from Brahmagupta's Khandakhadyaka. The "old Suryasiddhanta" summarized by Varaha Mihira is also based on the Midnight System.

In both systems the fundamental return period of all planets is the *Ma-hayuga* or fourfold yuga, containing

$$4\,320\,000$$

sidereal years. It is divided into 4 "quarteryugas" of equal duration. The last quarteryuga, in which we now live, is the *Kaliyuga*. In the Sunrise System the Kaliyuga begins at sunrise on

Friday, February 18, 3102 B.C.

and in the Midnight System it begins 6 hours earlier, at midnight. At this moment, and also at the beginning and end of the Mahayuga, all planets (including sun and moon) were supposed to have mean longitude zero. This implies that the numbers of revolution of the planets in a Mahayuga must be multiples of 4. The following table contains these numbers of revolutions and also the positions of the planetary apogees and the numbers of revolution of the moon's apogee and node.

Table 1. Revolutions in a Mahayuga

Planet	Midnight System		Sunrise System	
	Revolutions	Apogee	Revolutions	Apogee
Sun	4320000	80°	4320000	78°
Moon	57753336		57753336	
Moon's Apogee	488219		488219	
Moon's Node	−232226		−232226	
Mercury	17937000	220°	17937020	210°
Venus	7022388	80°	7022388	90°
Mars	2296824	110°	2296824	118°
Jupiter	364220	160°	364224	180°
Saturn	146564	240°	146564	236°

It is seen that the two systems differ very little. The numbers of revolution differ only for Mercury and Jupiter. As Billard has shown in his book "L'astronomie indienne", the mean positions of the planets, calculated for the time of Aryabhata (A.D. 510) are in very good agreement with modern calculations. For the Sunrise System they are even better than for the Midnight System.

On the other hand, the mean motions of the planets in both systems are not very accurate. The reason for this is that in reality (according to modern calculations) the mean longitudes were not at all zero at the beginning of the Kaliyuga. On February 18, 3102 B.C. the longitudes of Jupiter and Saturn differed by more than 40°, as I have shown in my paper "The Conjunction of 3102 B.C." in Centaurus 24, p. 117–131 (1980). As a consequence the longitudes, calculated by Aryabhata's theory, would considera-

bly deviate from reality after one or two centuries. Later astronomers had
to apply corrections either to the initial longitudes or to the numbers of
revolution or to both. These corrections enabled Billard to date many later
astronomers. See my paper "Two Treatises on Indian Astronomy", J. for
the History of Astronomy 11, p. 50–58 (1980).

Brahmagupta's System

The title of Brahmagupta's great astronomical treatise is "Brahma-
sphuta-siddhanta", which means something like "corrected doctrine of
Brahma". It was written in A. D. 628, about 120 years after the time of
Aryabhata.

Brahmagupta's fundamental period is the *Kalpa* of 1 000 Mahayugas, or
4 320 millions of years. At the beginning and end of this period all planets
are supposed to have mean longitude zero, and also the longitudes of their
apogees and nodes are supposed to be zero, which implies that their true
longitudes and latitudes are zero as well.

If Brahmagupta's numbers of revolution in a Kalpa are divided by
1 000, one obtains numbers of revolutions in a Mahayuga, which can be
compared with the numbers in Aryabhata's Sunrise System. The comparison is made in Table 2.

Table 2. Revolutions in a Mahayuga

Planet	Aryabhata	Brahmagupta
Sun	4 320 000	4 320 000
Moon	57 753 336	57 753 300
Moon's Apogee	488 219	488 105.858
Moon's Node	− 232 226	− 232 311.168
Mercury	17 937 020	17 936 998.984
Venus	7 022 388	7 022 389.492
Mars	2 296 824	2 296 828.522
Jupiter	364 224	364 226.455
Saturn	146 564	146 567.298

Note that the revolutions of Venus and Mercury are what we would call
heliocentric revolutions. In the terminology of the Hindu astronomers they
are revolutions of the *śighrocca* of Venus and Mercury, i. e. of the line
drawn from the centre of the "*śighra* epicycle" to the planet. Let me explain these notions by means of a drawing (Fig. 64).

I have made the drawing for Venus. The planet moves on a epicycle, the
"śighra epicycle". The centre *C* of this epicycle is made to move on an eccentric circle surrounding the earth *E*, the centre of the circle being at *B*. If
we complete the parallelogram *EBCD*, we see that the eccentric motion
can be replaced by an equivalent motion on a small epicycle with centre *D*
such that the radius *DC* is always parallel to the line *EA*. This small epicy-

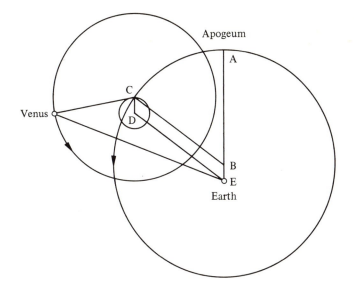

Fig. 64. The epicyclic motion of Venus

cle is called the "manda epicycle". Its centre D moves on a concentric circle (not shown in the figure) around the earth.

Aryabhata and Brahmagupta both use this "double epicycle theory". From our diagram it is clear that the motion of Venus according to this theory is equivalent to the motion on an epicycle carried by an eccentric circle.

From Table 2 one sees that Brahmagupta's numbers of revolution agree fairly well with those of Aryabhata. If we restrict ourselves to the integral parts, we see that for Venus and the superior planets Brahmagupta has applied the following corrections:

Venus	+1
Jupiter	+2
Saturn	+3
Mars	+4.

These corrections are very reasonable, for if one looks at Billard's diagram of deviations of Aryabhata's mean longitudes from modern calculations (Fig. 65), one sees that these deviations are practically zero for A. D. 510, and become negative in the course of time. Hence for the time of Brahmagupta positive corrections of the numbers of revolution are necessary in order to bring the theory into reasonable agreement with observations. The required correction is least for Venus, larger for Jupiter, still larger for Saturn, and largest for Mars. So we can understand Brahmagupta's corrections $+1, +2, +3$, and $+4$ very well.

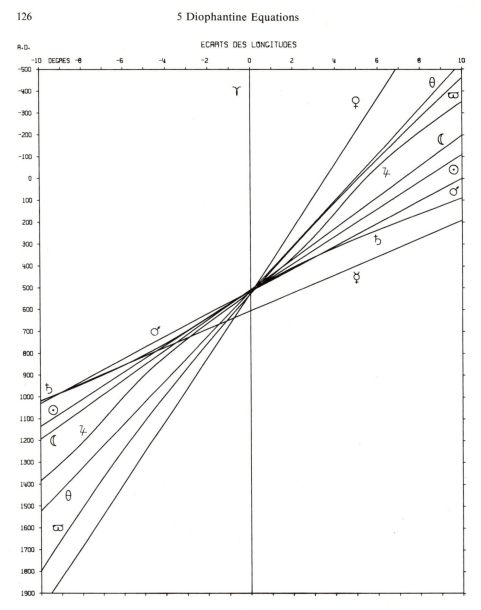

Fig. 65. Deviations of Aryabhata's Longitudes (from R. Billard: L'astronomie indienne, Paris 1971, Fig. 5). θ is the moon's node, $\bar{\omega}$ its apogee.

Even with these corrections, the mean motions are still far from reality. One or two centuries after Brahmagupta new corrections would be necessary. Obviously, Brahmagupta did not determine his mean motions empirically. I suppose he started with the assumption of an approximate conjunction of all planets on February 18, 3102 B.C., and he determined the numbers of revolution in such a way that for his own time the calculated longitudes would be in reasonable agreement with observed longitudes.

How did he reach this aim? What mathematical methods did he need?

At the beginning of the Kalpa, Brahmagupta assumed all mean longitudes of planets as well as the longitudes of their apogees and nodes to be zero. The current Kaliyuga was assumed to begin 456.7 Mahayugas after the beginning of the Kalpa. The time of Aryabhata was 3600 years after the beginning of the Kaliyuga. This is a short time as compared with the elapsed part of the Kalpa. We shall see that this short duration is essential for Brahmagupta's calculations.

The required numbers of revolutions had to satisfy three conditions:

1) The numbers of revolutions in a Magayuga should be approximately equal to those of Aryabhata, with small corrections as indicated.

2) The calculated positions of the planets in the time of Brahmagupta should be approximately in accordance with observations made at his own time or a little earlier.

3) The calculated positions of apogees and nodes should approximately coincide with those of Aryabhata.

We shall see that, in order to satisfy these conditions, one has to solve certain linear Diophantine equations.

Let us begin with the apogees and nodes, and next consider the mean motions of the planets.

The Motions of the Apogees and Nodes

In Aryabhata's Midnight System, the apogee of the sun is constant or nearly constant at 78°, and in his Sunrise System it is at 80°. Brahmagupta seems to have adopted the former value 78°, for in his system the sun's apogee is situated at 77°54' in A.D. 499.

In order to satisfy condition 1), Brahmagupta had to assume a small number of revolutions in a Kalpa, say less than 1000 revolutions. Actually his numbers of revolutions of apogees and nodes are all less than 900, as one sees from Table 3:

Table 3. Motion of Absides and Nodes in a Kalpa

Planet	Apogee	Node
Sun	480	
Mercury	332	−511
Venus	653	−893
Mars	292	−267
Jupiter	855	− 63
Saturn	41	−584

If the number of revolutions of the sun's apogee is less than 1000 in a Kalpa, it is less than 1 in a Mahayuga of 4320000 years. It follows that the motion in the 3 600 years from the beginning of the Kaliyuga to the time of Aryabhata is less than

$$\frac{3\,600}{4\,320\,000} \cdot 360° = 0.3 \text{ degrees}.$$

Therefore, if we take care that the apogee is near $78°$ at the beginning of the Kaliyuga, it will also be near $78°$ at the time of Aryabhata.

If the number of revolutions of the solar apogee in a Kalpa is x, the motion of the apogee from beginning Kalpa to beginning Kaliyuga is

$$\frac{4\,567}{10\,000} \cdot x \cdot 360°.$$

So, if we want to obtain a position near $78°$, we have to satisfy a Diophantine equation

$$\frac{4\,567}{10\,000} \cdot x \cdot 360 - 360\,y = g,$$

where g is approximately 78. Multiplying by 1 000, one obtains

(28) $4\,567 \cdot 36\,x - 360\,000\,y = 1\,000\,g.$

The left side of (28) is divisible by 36, so $1\,000\,g$ must also be divisible by 36. Putting

$$1\,000\,g = 36\,h$$

one obtains the Diophantine equation

(29) $4\,567\,x - 10\,000\,y = h$

where h must be approximately equal to

$$\frac{1\,000}{36} \cdot 78 = 2\,166 + 2/3.$$

Let us first try the next smaller integer, $h = 2\,166$. We have to solve the Diophantine equation

(30) $4\,567\,x - 10\,000\,y = 2\,166.$

By applying the "pulverizer", one obtains the smallest positive solution

$$x = 3\,898, \quad y = 1\,780.$$

I suppose Brahmagupta wanted positive solutions, for the motions of his apogees are all positive. If x is positive, the longitude of the apogee increases in the course of time, so if we want to have it near 78° for A.D. 499, we have to make it slightly less than 78° at beginning Kaliyuga. This means: we may diminish the value $h = 2166$, but not increase it. Now every time the constant h on the right side of (29) is diminished by one unit, the value of x is increased by 1097, because the smallest positive solution of

$$4567x - 10000y = -1$$

is $x = 1097$. So we may, starting with $h = 2166$ and $x = 3898$, diminish h by n units and at the same time add $n \cdot 1097$ to x. We may choose n in such a way that

$$x' = 3898 + n \cdot 1097$$

just exceeds 10000. In our case, n becomes 6, and x' becomes

$$x' = 3898 + 6 \times 1097 = 10480.$$

Next, x' can be diminished by 10000 and the corresponding y' by 4567. Thus one obtains the solution

$$x = 480, \quad y = 219, \quad h = 2160,$$

which Brahmagupta has adopted. The corresponding longitude of the apogee at beginning Kaliyuga is

$$g = \frac{36}{1000} h = 77 \cdot 760 \text{ degrees}.$$

The motion of the apogee during the 3600 years from beginning Kaliyuga to A.D. 499 is

$$\frac{3600 \times 480 \times 360}{4320000000} = 0.144 \text{ degrees},$$

so the longitude of the apogee in A.D. 499 is

$$77.760 + 0.144 = 77.904 \text{ degrees}$$

or 77° 54′, which is indeed quite near to 78°.

The same method can be applied to the apogees of the five "star-planets". I suppose that Brahmagupta applied this or a similar method. The values of x thus obtained are in any case less than 1097, because x is augmented by $n \cdot 1097$ until it just exceeds 10000, and next 10000 is subtracted. In fact, Brahmagupta's values are all less than 900 (see Table 3).

For the nodes, the method is just the same, the only difference being that the motions of the nodes are all assumed to be negative (see again Table 3).

The Motion of the Planets

For the planets the situation is more complicated, but the method still works. Let us divide the required number x into two parts

$$(31) \qquad\qquad x = 1000q + s.$$

Here q is the approximate number of revolutions in a Mahayuga, and s is a correction term less than 1000. To obtain q, Brahmagupta took the numbers of Aryabhata's Sunrise System, with corrections

$$+1, \quad +2, \quad +3, \quad +4$$

for Venus, Jupiter, Saturn, and Mars respectively, as we have seen earlier. I shall leave out of account the elusive planet Mercury.

As Billard has shown, Brahmagupta's numbers x were determined in such a way that the longitudes derived from his theory agreed approximately with observations made in the middle part of the sixth century, that is, more than 50 years before the time of Brahmagupta. For his own time, his longitudes are less good, as one sees from Billard's diagram reproduced here (Fig. 66). In the diagram the deviations of Brahmagupta's mean longitudes from modern tables are plotted as functions of time. I have drawn a horizontal line corresponding to the year A. D. 628, in which Brahmagupta composed his Siddhanta. Whereas the error for Venus is nearly zero, the error for Saturn amounts to about 2°, and for Jupiter, Mars, the sun, and the moon to about 1°.

Let x be the number of revolutions of any planet in a Mahayuga, and g the mean longitude at beginning Kaliyuga. Then we have, as before, an Eq. (29):

$$(32) \qquad\qquad 4567x - 10000y = h$$

with

$$(33) \qquad\qquad 36h = 1000g.$$

In the case of the apogees, the solution of this Diophantine equation was easy, because x was supposed to be small. In the case of the planets, x has to be of the form (31):

$$x = 1000q + s,$$

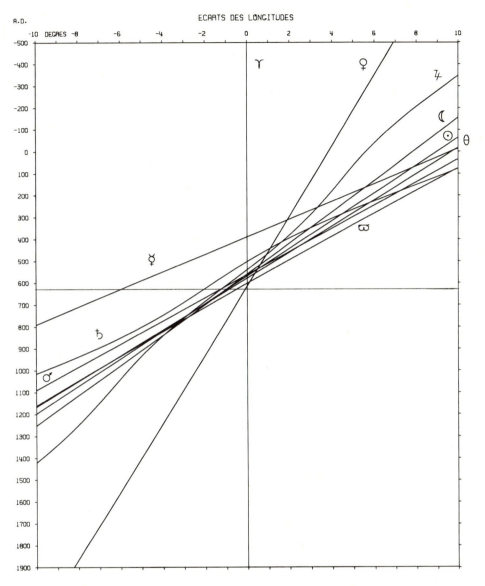

Fig. 66. Deviations of Brahmagupta's Longitudes (from R. Billard: L'astronomie indienne, Fig. 9). θ is the moon's node, $\bar{\omega}$ its perigee.

where q is given, and s has to be less than 1 000. Moreover, the longitude g at beginning Kaliyuga is required to be nearly equal to zero, because in Aryabhata's system this longitude is exactly zero. Brahmagupta managed, in all cases, to find solutions of (32) satisfying the two conditions for s and g just specified. For this, he needed the pulverizer.

The Influence of Hellenistic Ideas

As we have seen, the solution of linear Diophantine equations is an indispensable tool in the astronomical theory of Brahmagupta, and indeed in any theory in which all longitudes of planets, apogees and nodes are supposed to be zero at the beginning of the Kalpa.

Aryabhata's theory is based on the Mahayuga and not on the Kalpa. Still, in verse 3 of the first part of the Aryabhatiya, a Kalpa of 1008 Mahayugas is mentioned. Also the fact that Aryabhata has included in his treatise a description of the pulverizer seems to indicate that already in his time astronomical theories based on the Kalpa system existed, in which the pulverizer was used in order to determine the number of revolution in a Kalpa.

The genesis of these astronomical systems lies hidden in darkness, but still a few facts may be mentioned which shed a dim light on this genesis.

First, I may note that a Kalpa or "Day of Brahman", consisting of 1000 Mahayugas, is mentioned in a passage on cosmic periods occurring in the "Law of Manu" as well as in the Mahabharata. According to Bühler[10], the common source of the two closely connected passages existed already in the second century A. D. The Mahayuga was supposed to consist of 4 parts, and the duration of the last part, the Kaliyuga, was supposed to be one tenth of the Mahayuga, or 432000 years. The figure

$$432000 = 2 \times 60^3$$

is almost certainly of Babylonian origin. It also occurs in the fragments of the Babylonian priest and astrologer Berossos (about -300). For more details see my paper "The Great Year in Greek, Persian and Hindu Astronomy", Archive for History of Exact Sciences 18, p. 359–384 (1978).

The passage in The Law of Manu ends thus:

To whatever course of action the Lord first appointed each (kind of beings), that alone it has spontaneously adopted in each successive creation. Whatever he assigned to each at the (first) creation, noxiousness or harmlessness, gentleness or ferocity, virtue or sin, truth or falsehood, that clung (afterwards) spontaneously to it.

Here we have a fatalistic doctrine of endless repetition of all human actions. The same doctrine is also found in Greek sources. A fragment of Aristotle's pupil Eudemos quoted by Simplikios[11] says:

If one believes the Pythagoreans, I too shall stand here again in future, holding this little stick in my hand and telling you myths, and you will sit here before me just as you are sitting now.

According to Nemesios, bishop of Nemesa (about $+400$), the Stoic philosophers held the same opinion concerning eternal return as the Pythagoreans. "There will be a conflagration caused by the planets" so they taught "and afterwards the world will be recreated and everything will be as be-

---

10 G. Bühler: The Law of Manu. Sacred Books of the East, Vol. 25, p. lxxxiii.
11 Simplikios: Commentary to Aristotle's Physics, ed. Diels, p. 732.

fore: there will be a Plato and a Socrates, and an exact repetition of all motions in the sky and on earth."

The same idea of annihilation followed by a new creation is also found in the Mahabharata and in a hymn of the Magi summarized by Dion Chrysostomos (about +100). On this hymn see Bidez et Cumont: Les mages hellénisés I, p. 91-97, and II, p. 142-153.

Conjunctions of all planets at the beginning and end of a large cosmic period were a common idea in the Hellenistic period. Berossos, a priest of the Babylonian god Bel, who came to Greece about 300 B.C. and founded a school of astrology on the island Kos, tells us that a conflagration will take place when all planets come together in Cancer, and a deluge when they come together in Capricorn. In Persian sources[12] a deluge is said to take place whenever the planets come together in the space between Pisces 27° and Aries 1°. The last time this happened was in February, 3102 B.C. Aryabhata and Brahmagupta also assumed an approximate conjunction of all planets in February of the same year, only one day later than in the Persian system.

The astronomical systems of Aryabhata and Brahmagupta are based on the epicycle hypothesis. This hypothesis was fully developed by Apollonios of Perge (about 200 B.C.). So in this respect too Hindu astronomy is just a continuation of Hellenistic astronomy. Also, Hindu astrology is clearly derived from Hellenistic astrology.

We have seen that the pulverizer is an indispensable tool in the theory of Brahmagupta, and that it was known already to Aryabhata. We have also seen that many essential ideas in Hindu astronomy have their roots in Hellenistic astronomy and astrology. Hence it is reasonable to conjecture that the pulverizer was already known in the Hellenistic age.

It is true that the solution of linear Diophantine equations is not found in any Greek text. On the other hand, the method is based on the Euclidean algorithm. As we shall see presently, the Greeks were able to solve much more difficult number-theoretical problems by means of the Euclidean algorithm, namely problems connected with the rational approximation of irrational ratios and with "Pell's Equation". Excellent Greek mathematicians like Archimedes or Apollonios would have no difficulty in solving linear Diophantine equations by means of the Euclidean algorithm.

Another argument in favour of the hypothesis of a Greek or even pre-Greek origin of the solution of linear Diophantine equations is the fact that the Chinese Remainder Problem is mentioned in Chinese, Indian, and Byzantine sources. As we have seen, our Byzantine source has just the same numerical values of the divisors and remainders as Sun Tzu. Brahmagupta's numerical values are different, but the kind of problems is just the same, and Bhaskara II solves problems of this kind by means of the pulverizer.

---

12 E.S. Kennedy and B.L. van der Waerden: The World-Year of the Persians, J. Amer. Oriental Soc. 83, p. 315-327 (1963).

For these reasons I believe that in an ancient Greek source or in a pre-Greek tradition problems giving rise to linear Diophantine equations were treated systematically by means of the Euclidean algorithm, and that this ancient tradition was the common source of all later treatments. This hypothesis would explain the similarities between the texts of Brahmagupta, Sun Tzu, and Isaac Argyros concerning the "Chinese Remainder Theorem".

Part B

Pell's Equation

The Diophantine equation

$$(1) \qquad\qquad x^2 = D y^2 + 1$$

is the so-called "Equation of Pell". This name was coined by Euler. Although Pell has little to do with this equation[13], I feel we should retain this name. Names are just names, and nothing else. We also speak of the "Theorem of Pythagoras", of "Dirichlet's Principle", and of the "Laplace Transformation". If we use these traditional names, everyone familiar with the subject knows what is meant.

Euler solved equations of type (1) by means of "continued fractions". We shall see that the Greeks obtained solutions of (1) for special values of D by an equivalent method based on the Euclidean algorithm.

The Equation $x^2 = 2y^2 \pm 1$

In the "Republic" 546 C, Plato calls the number 7 the "rational diagonal", and from the commentaries we know that this number corresponds to the "rational side" 5. In fact, the ratio $7:5$ is a pretty good approximation to the ratio of the diagonal to the side of a square.

In his commentary to this passage, Proklos gives a definition of "side-and diagonal-numbers" as follows:

As the source of all numbers, unity is potentially a side as well as a diagonal. Now let two units be taken: one side unit and one diagonal unit. Then a new side is formed by adding the diagonal unit to the side unit, and a new diagonal by adding twice the side unit to the diagonal unit (Proklos, Commentary to the Republic, Vol. II, p. 24–25 in the edition of Kroll).

---

13 See L. E. Dickson: History of the Theory of Numbers II, p. 341.

Starting with a pair of units and applying the rule indicated by Proklos, one obtains in a first step the "side-number"

$$s = 1 + 1 = 2$$

and the corresponding "diagonal-number"

$$d = 2 \times 1 + 1 = 3.$$

Repeating this process according to the formula

$$s' = s + d$$
(2)
$$d' = 2s + d$$

one obtains $s' = 5$ and $d' = 7$, the pair mentioned by Plato. And so on.

If s_n and d_n are the n-th side-number and diagonal-number, we have

$$s_1 = d_1 = 1$$

and

(3)
$$s_{n+1} = s_n + d_n$$
$$d_{n+1} = 2s_n + d_n$$

The same recursion rule is also mentioned by Theon of Smyrna in his "Expositio rerum mathematicorum" on p. 43 in the edition of Hiller. Proklos ascribes the formation of side- and diagonal-numbers to the Pythagoreans.

The ratio $d_n : s_n$ is an increasingly accurate approximation to the ratio $d : s$ of the diagonal to the side of a square. This follows from the equality

(4)
$$d_n^2 = 2 s_n^2 \pm 1$$

which is also mentioned by Proklos. Dividing both sides by s_n^2, one obtains

$$(d_n / s_n)^2 = 2 \pm (1/s_n)^2 ,$$

so the square of d_n / s_n is nearly equal to 2.

The text of Proklos implies that the Pythagoreans proved (4) by means of Euclid's proposition II 10. As we have seen in Chapter 3, this proposition can be transcribed by the identity

(5)
$$(2s + d)^2 + d^2 = 2s^2 + 2(s + d)^2.$$

Proklos says that the Pythagoreans used II 10 to prove an "elegant theorem", namely: *If s and d are side and diagonal of a square, then s + d and*

$2s+d$ *are side and diagonal of another square.* In Chapter 3 we have seen that it is indeed easy to derive this theorem from the identity (5).

The linear transformation (2) is closely connected with the Euclidean algorithm. In fact, if we solve (2) for s and d, we obtain

$$s = d' - s'$$
$$d = s' - s.$$

Hence, if d' and s' are diagonal and side of a square and if we apply the Euclidean method of alternate subtraction, we obtain in two steps

$$(d',s') \to (s,s') \to (s,d).$$

This means: the transformation (2), applied to side and diagonal of a square, is just the inverse of the Euclidean algorithm, applied to the diagonal d' and side s' of a larger square.

Of course, the identity (5) is not only valid for line segments, but also for numbers. If we replace s and d by the side- and diagonal-numbers s_n and d_n, we obtain

$$(2s_n + d_n)^2 + d_n{}^2 = 2s_n{}^2 + 2(s_n + d_n)^2$$

or

(6) $$d_{n+1}{}^2 + d_n{}^2 = 2s_n{}^2 + 2s_{n+1}{}^2.$$

This equality is also mentioned by Proklos, so we may safely conclude that the Pythagoreans knew (6) and proved it by means of II 10.

Now we come to the proof of (4). Obviously, (4) holds for $s_1 = d_1 = 1$. But if (4) holds for a certain value of n, it follows from (6) that (4) also holds for $n+1$ with the opposite sign of the term ± 1. Hence (4) holds for arbitrary n, with alternating signs of the term ± 1.

This proof, which can easily be reconstructed from the text of Proklos, is the earliest known example of a proof by "complete induction"[14] or "passage from n to $n+1$". The Pythagorean who found the recursion rule (3) and who proved (4) must have been an excellent mathematician!

Periodicity in the Euclidean Algorithm

We have seen that the Euclidean algorithm, applied to the diagonal and side of a square, after two steps produces the side and diagonal of another square. If the process is repeated, it goes on infinitely.

Right at the beginning of book 10 of Euclid's Elements, we find two very remarkable theorems X1 and X2 on the Euclidean algorithm. The second of these says:

---

14 For the history of this method of proof see H. Freudenthal: Zur Geschichte der vollständigen Induktion, Archives internationales d'histoire des sciences 22, p. 17–37 (1953).

X2. If the smaller of two unequal magnitudes is continually subtracted from the larger, and if that what is left never measures the one before it, the magnitudes will be incommensurable.

This criterion for incommensurability is never used in the sequel. In the course of book 10, several proofs of incommensurability are given, but none of these proofs is based on the criterion X2. I suppose that this criterion is a residue from an earlier theory of irrational ratios.

Can we say more about this earlier theory?

In Plato's dialogue Theaitetos the protagonist Theaitetos gives an account of a lecture given by Theodoros of Kyrene, in which the latter proved that the sides of squares of areas 3,5,6, etc. (square) foot are not commensurable with the unit of length. He took these squares singly, one after the other, and he stopped at or just before the 17-foot square. Can we make a reasonable guess about his method of proof?

I shall denote by $w_2, w_3, \ldots$ the sides of the squares having areas of $2, 3, \ldots$ square feet. In particular, w_2 denotes the diagonal of the unit square. The unit of length (the "foot" in Plato's text) will be denoted by e.

Theodoros began his exposition with w_3, possibly because the incommensurability of w_2 and e had already been proved by the Pythagoreans. From Plato's text we may conclude that Theaitetos formulated a general theorem saying that w_n is commensurable with e if and only if n is a square. It seems that Theodoros did not know the general method of Theaitetos, and instead used special methods for individual cases such as w_3, w_5, etc.

A very plausible hypothesis due to Zeuthen says that Theodoros used the criterion X2. We have seen that the Euclidean algorithm, applied to w_2 and e, i.e. to the diagonal d' and the side s' of a square, yields after two steps the diagonal d and the side s of a smaller square, which implies that the "reciprocal subtraction" never ends. So Theodoros might well have tried to apply the same algorithm to w_3 and e, and to see whether the ratio $w_3 : e$ repeats itself after a few steps. Actually it repeats itself after 3 steps. This can be proved as follows, by a reasoning well within the power of the early Greek geometers.

Put $w_3 = w$. Then we have

$$w^2 = 3 e^2.$$

Subtract e from w: the remainder is $w - e$. Instead of the original line segments e and w we now have the segments e and $w - e$. In continuing the procedure, we can replace e and $w - e$ by any two segments having the same ratio. Now

(7) $$e : (w - e) = (w + e) : 2e$$

because the product of the inner terms equals that of the outer terms:

$$(w - e)(w + e) = w^2 - e^2 = 3 e^2 - e^2 = 2 e^2.$$

Next we apply the same method to the line segments $w+e$ and $2e$. Subtracting $2e$ from $w+e$, we obtain $w-e$. The line segments $2e$ and $w-e$ can again be replaced by two segments having the same ratio:

(8) $$2e:(w-e)=(w+e):e.$$

Once more, the product of the inner terms equals the product of the outer terms. If we now take $w+e$ and e, and subtract the latter from the former, we obtain w and e: the segments from which we started. The ratio $w:e$ is repeated, so the process is periodic and will never end. Hence w_3 and e are incommensurable.

It is not certain that Theodoros used the Euclidean algorithm. Another possibility was indicated by W. R. Knorr in his book "The Evolution of the Euclidean Elements", Dordrecht 1975. However, I feel that Zeuthen's hypothesis is plausible, for two reasons. First, the Pythagoreans had already demonstrated that the Euclidean algorithm, applied to w_2 and e, is periodic, and Theodoros only had to apply the same method to w_3, w_5, etc. Secondly, if Theodoros proceeded like this, we can understand why Euclid included the historically important proposition X2 in his Elements.

Reciprocal Subtraction

The Greek word for the Euclidean algorithm is Antanairesis, which means reciprocal subtraction. One starts with a pair of numbers (x,y) or with a pair of line segments (u,v), and one continually subtracts the smaller one from the larger one. Either the process ends with two equal terms, or it goes on infinitely.

In Chapter 2 I have illustrated the process by an example from the Chinese "Nine Chapters". According to the Chinese text, the numbers x and y are "laid out" on the counting board. If x is larger than y, the first operation is, to subtract y from x. I shall call this operation S:

S: $$(x,y) \rightarrow (x-y,y).$$

But if y is larger than x, the first operation is, to subtract x from y:

T: $$(x,y) \rightarrow (x,y-x).$$

If we apply these two operations, one after the other, to the diagonal d' and side s' of a square, we obtain after two steps the side s and diagonal d of another square as we have seen.

The inverse operations are additions

A: $$(x,y) \rightarrow (x+y,y)$$
and
B: $$(x,y) \rightarrow (x,x+y).$$

If we apply these two operations to the side s and diagonal d of a square, we obtain after two steps

$$(s,d) \to (s,s+d) \to (2s+d,s+d).$$

This is the operation by which the Pythagoreans obtained their side- and diagonal-numbers.

Now let us see whether the same method can be applied to a pair of line segments u' and v' which are proportional to w_3 and e. If we apply the operator STS, we obtain a pair of line segments (u,v) proportional to (u',v'), as we have seen. The inverse operator is ABA:

$$(u,v) \to (u+v,v) \to (u+v,u+2v) \to (2u+3v,u+2v).$$

So we obtain the result:

If u and v are line segments satisfying the equation

(9) $$u^2 = 3v^2$$

then

(10)
$$u' = 2u+3v$$
$$v' = u+2v$$

are again line segments satisfying the same equation (9).

This theorem can be verified by an elementary calculation:

(11) $$(2u+3v)^2 - 3(u+2v)^2 = u^2 - 3v^2.$$

The two squares on the left can be evaluated by Euclid's proposition II 4. Of course, one can also avoid negative terms and write (11) in the Euclidean fashion as

$$(2u+3v) + 3v^2 = 3(u+2v)^2 + u^2.$$

Now let us apply the linear transformation (10) and the identity (11) to the problem of solving Pell's equation for $D=3$.

The Equations $x^2 = 3y^2 + 1$ and $x^2 = 3y^2 - 2$

In the case $D=2$ we have seen that the Pythagoreans obtained a sequence of approximations to the ratio $w_2 : e$ by starting with $1:1$ and applying the recursive process (3). In the case $D=3$ one can start with the approximation $2:1$ and apply the linear transformation (10) repeatedly, thus obtaining

(12)
$$x_{n+1} = 2x_n + 3y_n$$
$$y_{n+1} = x_n + 2y_n.$$

The pair $(2,1)$, from which we started, satisfies the equation

$$x^2 = 3y + 1.$$

Now suppose that we have, for a certain value of n

(13) $x_n^2 = 3y_n^2 + 1.$

Applying the identity (11), one obtains

$$(2x_n + 3y_n)^2 - 3(x_n + 2y_n)^2 = 1$$

or

$$x_{n+1}^2 = 3y_{n+1}^2 + 1.$$

Thus, by complete induction, we see that (13) holds for all n.

Because x_n^2 is larger that $3y_n^2$, the ratio $x_n : y_n$ is larger than the ratio $w_3 : e$. To obtain an approximation on the lower side, we may start with the pair $(1,1)$ and apply the same recursive process (12). We thus obtain pairs of numbers (s_n, t_n) satisfying the condition

$$s_n^2 = 3t_n^2 - 2.$$

The computation of the pairs (x_n, y_n) and (s_n, t_n) yields

$n = 1$	2	3	4	5	6
$x_n = 2$	7	26	97	362	1351
$y_n = 1$	4	15	56	209	780
$s_n = 1$	5	19	71	265	
$t_n = 1$	3	11	41	153	

The pairs $(1\,351, 780)$ and $(265,153)$ were known to Archimedes, as we shall see presently.

Our solution of Pell's equation (1) for $D = 2$ was based on the identity (5). Just so, our solution for $D = 3$ was based on the identity (11). Both identities (5) and (11) are special cases of a more general identity

(14) $(ax + Dcy)^2 - D(cx + ay)^2 = (a^2 - Dc^2)(x^2 - Dy^2).$

This identity, which can easily be verified, was used by Brahmagupta and by Euler and Lagrange for solving Pell's equation. Brahmagupta's method will be discussed later in this chapter.

Archimedes' Upper and Lower Limits for w_3

In his extremely interesting paper "On the Measurement of the Circle" Archimedes proves that the circum ference of a circle is less than $3 + 1/7$ times, but larger than $3 + 10/71$ times the diameter. He starts with a rectangular triangle in which one angle is $1/3$ of a right angle. If the opposite side is taken as a unit e, the hypotenuse is $2e$, and the third side is $w = w_3$, for its square is $(2e)^2 - e^2 = 3e^2$. Archimedes now asserts that the ratio $w:e$ lies between the limits

$$265:153 \quad \text{and} \quad 1351:780.$$

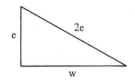

Fig. 67. The triangle from which Archimedes started

In modern terminology, we can transcribe this as

$$\frac{265}{153} < \sqrt{3} < \frac{1351}{780} .$$

It is easy to verify that these inequalities are correct, for the numbers 265 and 153 satisfy the equation

(15) $$x^2 = 3y^2 - 2$$

and the other two numbers satisfy

(16) $$x^2 = 3y^2 + 1 .$$

How did Archimedes obtain his limits for the ratio $w:e$? Historians of mathematics have made several conjectures to answer this question. In my opinion the most interesting papers are:
1) P. Tannery: Sur la mesure du cercle d'Archimède, Mémoires scientifiques I, p. 246–253.
2) C. Müller: Wie fand Archimedes die von ihm gegebenen Näherungswerte von $\sqrt{3}$? Quellen und Studien zur Geschichte der Mathematik B 2, p. 281–285.
3) O. Toeplitz: Bemerkungen zu der vorstehenden Arbeit von Conrad Müller, same journal, p. 286–290.
4) K. Vogel: Jahresbericht der Deutschen Mathematiker-Vereinigung 41, p. 5–8.

Note added in proof. See also W. R. Knorr: Archimedes and the Measurement of the Circle, Archive for History of Exact Sciences 15, p. 115–140 (1976).

One possibility to explain the Archimedean limits would be to suppose that he followed the procedure explained in the preceding section: Starting with the special solutions (1,1) and (2,1) of (15) and (16) respectively, he might have applied the recursive process (12). This explanation was given by Tannery in his paper 1).

In paper 2), Müller has derived a general rule for obtaining lower and upper bounds for square roots $\sqrt{D}$, which leads in three steps to the bounds for $\sqrt{3}$ given by Archimedes. The principle of Müller's method is: *If α and β are both upper or both lower bounds for $\sqrt{D}$, then*

$$\gamma = (\alpha\beta + D)/(\alpha + \beta)$$

is an upper bound. If α is a lower and β an upper bound, γ is a lower bound.

In his paper 3), Toeplitz has shown that this rule is an immediate consequence of Brahmagupta's identity (14). Thus, no matter whether one accepts Tannery's explanation or Müller's, one always comes back to Brahmagupta's identity for $D=3$.

We have seen that the recursive process (12) leads to pairs (x_n,y_n) and (s_n,t_n) satisfying (16) and (15) respectively. Let the pairs of the first sequence be called A_n and those of the second B_n. Toeplitz has shown: If Brahmagupta's identity for $D=3$ is applied to two pairs A_m and A_n, one obtains the pair A_{m+n}, and if it is applied to A_m and B_n, one obtains B_{m+n}. This means:

$$x_m x_n + D y_m y_n = x_{m+n}$$
$$x_m y_n + y_m x_n = y_{m+n}$$

and

$$x_m s_n + D y_m t_n = s_{m+n}$$
$$x_m t_n + s_m y_n = t_{m+n}.$$

These formulae yield a rapid method for computing Archimedes' pairs

$$A_6 = (1\,351,780) \quad \text{and} \quad B_5 = (265,153)$$

as follows: from A_1 and A_1 one gets A_2, from A_2 and A_2 one gets A_4, from A_2 and A_4 one gets A_6, and from A_4 and B_1 one gets B_5. It is not necessary to compute intermediate pairs.

Let me now summarize. If we assume that the ancient Greek geometers systematically used the Euclidean algorithm to obtain rational approximations of irrational ratios and to solve Pell's equation for special values of D, then we understand how the Pythagoreans came to the side- and diagonal-numbers,

we can see why Theodoros of Kyrene, in proving the incommensurability of certain sides of squares, gave separate proofs for each single square of area 3,5, etc.,

we can understand why Euclid included a criterion for incommensurability (X 2) in his Elements, although he did not need it in the sequel,

and we can see how Archimedes might have obtained his approximations for the ratio $w_3 : e$.

It is true that this hypothesis is not proved, but it explains several facts, and it is not disproved by any known fact. I feel that's all we can expect of a good hypothesis.

We shall see that the same hypothesis can also be used to elucidate the methods for solving Pell's equation we find in Indian sources.

Continued Fractions

The methods of Euler and Lagrange to solve Pell's equation are based on the development of $\sqrt{D}$ into a continued fraction. Starting with $\sqrt{D}$, one substracts 1 as often as possible until the remainder r is less than 1. Next one calculates $1/r$ and again substracts 1 as often as possible, and so on. In the case $D = 3$ one obtains

$$\sqrt{3}$$
$$\sqrt{3} - 1$$
$$1/(\sqrt{3} - 1) = (\sqrt{3} + 1)/2$$
$$(\sqrt{3} + 1)/2 - 1 = (\sqrt{3} - 1)/2$$
$$2/(\sqrt{3} - 1) = \sqrt{3} + 1$$
$$\sqrt{3}, \text{ etc.}$$

It is clear that this process is essentially the same as the "antanairesis" process, applied to $w = w_3$ and e, which we have discussed earlier. The single subtractions in the antanairesis process correspond to the single subtractions in the continued fraction development.

If, at a certain stage of the development, the remainder r is replaced by zero, one obtains an approximation. The approximations are alternately too small and too large. Thus, in the continued fraction development of $\sqrt{2}$, the first approximation 1 is too small, the second approximation $1 + \frac{1}{2} = 3/2$ is too large, etc. The successive approximations are just the quotients of the diagonal- and side-numbers.

The same process can also be used to obtain simple fractions as approximations of more complicated fractions. The Greeks have made ample use of this possibility. By way of an example, let us consider the calculation of the circumference of the circle by Archimedes.

Archimedes first states that the circumference of the circle is less than that of a circumscribed polygon and larger than that of an inscribed polygon. He calculates the perimeter of a circumscribed regular polygon of 96 sides, and he finds that the ratio of this perimeter to the diameter of the circle is less than

$$14688 : (4673 + 1/2).$$

Next he says: The former number is less than $3 + 1/7$ times the first, so that the perimeter of the circumscribed 96-gon, and a fortiori the circumference of the circle, is less than $3 + 1/7$ times the diameter.

The result 22/7 is a continued fraction approximation of $14688/(4673 + 1/2)$. It can easily be found by means of the Euclidean algorithm, as follows:

$$\begin{vmatrix} 14688 \\ 4673 + 1/2 \end{vmatrix} \text{3 subtractions} \begin{vmatrix} 667 + 1/2 \\ 4673 + 1/2 \end{vmatrix} \text{7 subtractions} \begin{vmatrix} 667 + 1/2 \\ 1 \end{vmatrix}$$

Replacing the numbers $667 + 1/2$ and 1 by 1 and 0, and working backwards, one obtains

$$\begin{vmatrix} 1 \\ 0 \end{vmatrix} \text{7 additions} \begin{vmatrix} 1 \\ 7 \end{vmatrix} \text{3 additions} \begin{vmatrix} 22 \\ 7 \end{vmatrix}$$

The lower limit $3 + 10/71$ can be obtained in the same way. I shall reproduce only the last part of the calculation

$$\begin{vmatrix} 10 \\ 7 \end{vmatrix} \text{7 additions} \begin{vmatrix} 10 \\ 71 \end{vmatrix} \text{3 additions} \begin{vmatrix} 223 \\ 71 \end{vmatrix}.$$

Still more instructive is a calculation made by Eratosthenes, who was an elder contemporary of Archimedes. According to Ptolemy (Almagest I,12), Eratosthenes estimated the arc between the extreme declinations of the sun to be 11/83 of a full circle, which means that the inclination of the zodiacal circle is about $23°51'20''$. Now I cannot believe that the fraction 11/83 is a direct result of an observation. No astronomer would ever divide a circle into 83 equal parts for the purpose of measuring culmination altitudes of the sun! However, it is quite possible that 11/83 was a result of a continued fraction approximation.

Suppose, for example, that Eratosthenes divided a quartercircle into 100 equal parts, and that he found that the arc between the highest and lowest altitude of the sun at noon is just 53 of these parts. He may also have divided the quarter-circle into 90 degrees and estimated the arc as $47 + 7/10$ degrees: the result would exactly be the same. Now comparing the 53 parts with the 400 parts of a full circle, he might perform an Antanairesis

$$\begin{vmatrix} 53 \\ 400 \end{vmatrix} \text{ 7 subtractions } \begin{vmatrix} 53 & 24 & 24 \\ 29 & 29 & 5 \end{vmatrix} \text{ 4 subtractions } \begin{vmatrix} 4 & 4 \\ 5 & 1 \end{vmatrix}$$

Replacing 4 and 1 by 1 and 0 and working backwards he would obtain

$$\begin{vmatrix} 1 & 1 \\ 0 & 1 \end{vmatrix} \text{ 4 additions } \begin{vmatrix} 5 & 5 & 11 \\ 1 & 6 & 6 \end{vmatrix} \text{ 7 additions } \begin{vmatrix} 11 \\ 83 \end{vmatrix}$$

The possibility of explaining the fraction 11/83 as the result of a continued fraction approximation was brought to my attention by Dennis Rawlins (San Diego, California).

The Equation $x^2 = Dy^2 \pm 1$ for Non-square D

Lagrange has proved (Oeuvres I, p. 671–731) that the same algorithm we have explained for $D=2$ and $D=3$ always leads to a periodic antanairesis, provided D is not a square. This means: if we start with two line segments u' and v' satisfying the equation

(17)
$$u'^2 = Dv'^2$$

the antanairesis leads, after a finite number of steps, to line segments u and v proportional to u' and v'. As before, we can invert the process and express u' and v' in terms of u and v:

(18)
$$u' = au + bv$$
$$v' = cu + dv$$

where a,b,c,d are positive integers. In special cases such as $D=2,3,5,\ldots$ one can determine the coefficients a,b,c,d in (18), and one sees that in all those cases the linear transformation (18) has the special form

(19)
$$u' = au + cDv$$
$$v' = cu + \quad av$$

with

(20)
$$a^2 - c^2 D = \pm 1.$$

With modern methods, it is easy to prove that (19) and (20) hold for any non-square D. I don't know whether the Greeks were able to prove this, but in any case they were able to determine the coefficients a and c and to verify (20) in every particular case.

Now if one starts with a particular solution (x,y) of the equation

(21)
$$x^2 = Dy^2 \pm 1,$$

for instance with the trivial solution (1,0), and if one applies the linear transformation (19) to x and y:

$$x' = ax + cDy$$
$$y' = cx + ay$$

one obtains a new pair (x',y'), which automatically satisfies the equation (21). The proof, by Brahmagupta's identity (14), is easy:

$$x'^2 - Dy'^2 = (ax + cDy)^2 - D(cx + ay)^2$$
$$= (a^2 - Dc^2)(x^2 - Dy^2)$$
$$= (\pm 1) \cdot (\pm 1) = \pm 1.$$

Thus, one is led to the recursive process

(22)
$$x_{n+1} = ax_n + cDy_n$$
$$y_{n+1} = cx_n + ay_n.$$

I suppose that this method of solving (21) was known to the Greeks of the Hellenistic age. We know that Archimedes not only obtained solutions of (15) and (16) for $D = 3$, but that he also occupied himself with Pell's equation for higher values of D, for his famous "Cattle Problem" leads to an equation

(23) $$x^2 = Dy^2 + 1$$

with

$$D = 2 \times 3 \times 7 \times 11 \times 29 \times 353$$
$$= 4\,729\,494.$$

See for this subject L. E. Dickson: History of the Theory of Numbers II, p. 342–345.

Brahmagupta's Method

In Brahmagupta's astronomical treatise "Brahma-sphuta-siddhanta" the eighteenth book is concerned with algebra and arithmetic, including the solution of linear Diophantine equations and of Pell's equation in the generalized form

(24) $$x^2 = Dy^2 + f$$

for a given additive f.

Brahmagupta calls D the "multiplier", x the "last root", and y the "first root". Bhaskara II, who elaborated Brahmagupta's theory in his Siddhantasiromani (written 1150 A. D.), calls x and y the "greatest" and "least" roots.

The "additive" f in (24) may also be negative. Brahmagupta freely uses negative numbers: he gives rules for their addition, subtraction, multiplication, and division (see H.T. Colebrooke, Algebra with Arithmetic and Mensuration from the Sanskrit of Brahmegupta and Bháscara, p. 339).

Right in the beginning of his exposition, Brahmagupta formulates a rule, by which solutions of (24) can be composed to form other solutions. His explanation is not easy to understand, but Bhaskara II gives a completely clear exposition. Starting with two solutions of (24), one for additive f and another for additive f', Bhaskara obtains a new solution for additive ff' thus:

The greatest and least roots are to be reciprocally multiplied crosswise, and the sum of the products to be taken for a least root. The product of the two (original) least roots being multiplied by the given multiplier, and the product of the greatest roots being added thereto, the sum is the corresponding greatest root, and the product of the additives will be the (new) additive (Colebrooke, p. 171).

In modern notation, this rule of composition may be formulated thus:

If (x,y) is a solution of (24) for additive f, and if (x',y') is a solution for additive f', then

(25)
$$x''=xx'+Dyy'$$
$$y''=xy'+\ yx'$$

is a solution for additive ff':

$$(xx'+Dyy')^2=D(xy'+yx')+ff'.$$

We can write this result as

$$(xx'+Dyy')^2-D(xy'+yx')^2=(x^2-Dy^2)(x'^2-Dy'^2),$$

which is just the identity (14).

Brahmagupta and Bhaskara also note that x and y may be divided by a common factor. The additive is then divided by the square of that factor.

Bhaskara's first example is the equation

(26) $x^2=8y^2+1.$

An obvious solution is $x=3, y=1$. Composing this solution with itself, one obtains

$$x'=17,\quad y'=6.$$

Composing this solution with the first, one obtains

$$x''=99,\quad y''=35$$

"and so on", says Bhaskara.

His second example is

(27) $x^2 = 11y^2 + 1$.

The pair $(3,1)$ belongs to the additive -2:

$$3^2 = 11 \times 1^2 - 2.$$

Composing this pair with itself, one obtains the pair $(20,6)$ with additive $+4$. Dividing by 2, one obtains a solution for additive $+1$:

$$10^2 = 11 \times 3^2 + 1.$$

Composing this solution with itself, Bhaskara obtains

$$199^2 = 11 \times 60^2 + 1.$$

"In like manner, an indefinite number of roots may be obtained", says Bhaskara.

It is clear that Brahmagupta's identity (14) is just a generalization of the Pythagorean identity (5), and that Bhaskara's method of obtaining an infinity of solutions of Pell's equation for $D=8$ and $D=11$ is a generalization of the Pythagorean construction of side- and diagonal-numbers. If one starts with the solution $(1,1)$ of the equation

$$x^2 = 2y^2 - 1$$

and applies Bhaskara's method repeatedly, one obtains the whole sequence of side- and diagonal-numbers.

Brahmagupta was mainly an astronomer. His astronomy is based on the Greek idea of explaining the motions of the planets by means of epicycles and eccenters. He uses trigonometrical methods, just as the Hellenistic astronomers did. In the work of the Hindu astronomers one finds Sanskrit words derived from Greek, such as "kendra" (distance from the centre, from Greek kentron) and "lipta" (minute of arc, from Greek lepton).

In Part A we have seen that Brahmagupta had a good reason for including the solution of linear Diophantine equations in his astronomical treatise. He and his predecessors needed these equations for determining the numbers of revolutions in the Kalpa system. But why should he be interested in the solution of Pell's equation?

The Greeks had a motive for occupying themselves with Pell's equation. The Pythagoreans wanted to find rational approximations for what we call $\sqrt{2}$, and they found the "Side- and Diagonal-Numbers". Archimedes needed approximations for $\sqrt{3}$, and he used for this purpose solutions of the equations

$$x^2 = 3y^2 + 1 \quad \text{and} \quad x^2 = 3y^2 - 2.$$

We also have seen that the methods for obtaining these solutions are closely connected with the Euclidean algorithm, applied to irrational ratios of line segments. But why should Brahmagupta explain these methods, stripped of their original context, in an astronomical treatise? I can find only one explanation: he followed an earlier tradition ultimately derived from Greek sources.

The Cyclic Method

We have seen that Euler solved Pell's equation by means of continued fractions. From $\sqrt{D}$ he subtracted an integer such that the remainder r lies between 0 and 1, next he formed $1/r$ and again subtracted an integer, and so on.

Euler himself noted that the procedure can sometimes be abridged by using negative remainders (see Euler: Vollständige Anleitung zur Algebra II, Kap. 7, Art. 101). Following this suggestion, C.O. Selenius developed an "ideal method" for solving Pell's equation, in which the number of steps is minimized. See C.O. Selenius: Konstruktion und Theorie halbregelmäßiger Kettenbrüche, Acta Acad. Aboensis (Math.-Phys.) 22, p. 1–77 (1960).

Afterwards, Selenius discovered that his "ideal method" is equivalent to the *Chakravala* (= cyclic) method taught by Jayadeva and Bhaskara II. See for this method:

K. Shankar Shukla: Acarya Jayadeva, the Mathematician. Ganita 5, p. 1–20 (1954).

H.T. Colebrooke: Algebra with Arithmetic from the Sanskrit of Brahmegupta and Bháscara, London 1817, p. 175–178.

H.O. Selenius: Rationale of the Chakravāla Process of Jayadeva and Bhāskara II. Historia Math. 2, p. 167–184 (1975).

I shall now explain the cyclic method by means of an example taken from the Siddhantasiromani of Bhaskara II, namely the solution of the equation

$$(28) \qquad x^2 = 67y^2 + 1.$$

Bhaskara starts with the pair $(8,1)$ satisfying

$$(29) \qquad 8^2 = 67 \times 1^2 - 3$$

with "additive" -3, and he composes it with a pair $(s,1)$, with a suitable value of s, according to the composition rule of Brahmagupta. The pair $(s,1)$ satisfies the trivial equality

$$s^2 = 67 \times 1^2 + (s^2 - 67)$$

with additive $s^2 - 67$. The composition yields a pair

$$(8s + 67, 8 + s)$$

satisfying

(30) $(8s + 67)^2 = 67(8 + s)^2 - 3(s^2 - 67)$

with additive $-3(s^2 - 67)$.

Now s is determined in such a way that $8 + s$ is divisible by -3, and that $|s^2 - 67|$ is as small as possible. The solution is $s = 7$. So (30) reduces to

$$123^2 = 67 \times 15^2 + 3 \times 18.$$

Dividing both sides by 3^2, one obtains

(31) $41^2 = 67 \times 5^2 + 6.$

The pair $(41, 5)$ is now composed with another pair $(t, 1)$. The result is

$$(41t + 5 \times 67)^2 = 67(5t + 41)^2 + 6(t^2 - 67).$$

Now t is determined in such a way that $5t + 41$ is divisible by 6. This means: one has to solve the linear Diophantine equation

(32) $5t + 41 = 6z.$

The solution, found by the method of the pulverizer, is

$$t = 5 + 6m.$$

The value of t which minimizes $|t^2 - 67|$ is $t = 5$. So we obtain

$$540^2 = 67 \times 66^2 - 6 \times 42$$

or, if we divide both sides by 6^2,

(33) $90^2 = 67 \times 11^2 - 7$

with additive -7.

And so on. The next equation would be

(34) $221^2 = 67 \times 27^2 - 2$

with additive -2. The continuation of the same method would lead, in several steps, to a solution of (28). However, it is possible to reach the same goal by composing the pair (221,27) with itself. Thus one obtains the pair

$$221^2 + 67 \times 27^2 = 97\,684$$
$$2 \times 221 \times 27 \quad = 11\,934$$

and the equation

$$97\,684^2 = 67 \times 11\,934^2 + 4.$$

Dividing both sides by 2^2, one obtains the final solutions

(35) $$48\,842^2 = 67 \times 5\,967^2 + 1.$$

Comparison Between Greek and Hindu Methods

We have seen that the periodicity of the antanairesis for w_2 and e can be proved by means of the proportion

$$e:(w_2 - e) = (w_2 + e):e$$

and the periodicity for w_3 and e by the proportions

$$e:(w_3 - e) = (w_3 + e):2e$$
$$2e:(w_3 - e) = (w_3 + e):e.$$

Now let us apply the same method to the case $D = 67$. From $w = w_{67}$ we can subtract $8e$, thus obtaining the pair

$$(e, w - 8e).$$

Now we have

(36) $$e:(w - 8e) = (w + 8e):3e,$$

because the product of the inner terms is equal to the product of the outer terms:

(37) $$w^2 - 64e^2 = 3e^2.$$

The approximation $p_1 = 8$ was chosen in such a way that $|p_1^2 - D|$ is as small as possible. The equation (37) is equivalent to

$$D - 64 = 3$$

or to

$$8^2 - D \cdot 1^2 = -3,$$

so that the pair (8,1) corresponds to the "additive" -3, just as in the cyclic method.

Next we have to subtract from $w+8e$ a multiple of $3e$, thus obtaining

$$(w+8e)-n\cdot 3e=w-(3n-8)e=w-p_2e$$

and to choose $p_2=3n-8$ in such a way that $|p_2^2-D|$ is as small as possible. Now 8 was our first approximation p_1, so p_2 has to satisfy the condition

$$p_2=3n-p_1$$

or

(38) $p_1+p_2=3n.$

The minimum of $|p_2^2-D|$ under the restriction (38) is attained for $p_2=7$.

Now compare this with the cyclic method. In our explanation of this method, we have composed the pair $(p_1,1)=(8,1)$ with a pair $(s,1)$ to form a new pair

$$x''=8s+67$$
$$y''=8+s.$$

The number s was determined in such a way that $8+s=p_1+s$ is divisible by 3, and that $|s^2-67|$ is as small as possible. So s is just the same as p_2:

$$s=p_2=7.$$

We now have $x''=123$ and $y''=15$. Dividing by 3, one obtains the pair

$$(x_2,y_2)=(41,5)$$

with additive 6 according to (31).

The third step reveals a difference between the two methods, but the final result is the same. We have obtained the pair of line segments $3e$ and

$$w-p_2e=w-7e.$$

Their ratio is

(39) $3e:(w-7e)=(w+7e):6e,$

because the product of the inner terms is equal to the product of the outer terms:

$$w^2-49e^2=18e^2.$$

The factor 6 in (39) is equal to the additive 6 in equation (31).

The next step in the antanairesis process would be, to subtract from $w+7e$ a multiple of $6e$, thus obtaining

$$(w+7e)-n'\cdot 6e = w-(6n'-7)e = w-p_3 e$$

and to determine $p_3 = 6n'-7$ by the condition

$$|p_3^2 - D| = \min.$$

Now 7 was our p_2, so we have

(40) $$p_2 + p_3 = 6n'.$$

The minimum solution is

$$p_3 = 5, \quad n' = 2.$$

This is a very simple method to determine p_3. The Hindus have a more complicated method. They compose the solution

$$(x_2, y_2) = (41, 5)$$

with another pair $(t, 1)$ so as to obtain a pair

$$(41t + 5 \times 67, 5t + 41),$$

and they determine t by the condition (32):

$$5t + 41 = 6z.$$

This condition is automatically satisfied if we choose $t = p_3 = 5$ according to the condition (40). Quite generally, one can prove by complete induction that the condition

(41) $$p_n + p_{n+1} \equiv 0 \pmod{q_n}$$

ensures the divisibility of

$$x'' = x_n p_{n+1} + D y_n$$

and

$$y'' = x_n + p_{n+1} y_n$$

by q_n, and of the corresponding additive by q_n^2, so that the cyclic process can be continued. For more details and proofs I refer to my paper "Pell's equation in Greek and Hindu Mathematics", Russian Math. Surveys 31:5, p. 210–225 (1976).

Why did Jayadeva and Bhaskara II use such a complicated method to ensure the divisibility of the "roots" x'' and y'' by q_n? I suppose they did not known that the idea underlying the cyclic method is the periodicity of the Euclidean algorithm. If one applies this algorithm, one has to subtract from $w + 7e$ a multiple of $6e$ and not just a multiple of e. It seems that Jayadeva and Bhaskara II did not known this. I suppose they realized: If we compose the solution (x_2, y_2) with another solution $(p_3, 1)$ without taking care that $x_2 + p_3 y_2$ be divisible by q_2, we get higher and higher additives, and we never arrive at a solution for additive 1. To avoid this difficulty, they solved the linear Diophantine equation

$$x_2 + p_3 y_2 = q_2 z$$

for p_3 and z by means of the pulverizer. The idea to use the pulverizer in order to ensure divisibility by q_2 was very clever, but they could have avoided it by making $p_2 + p_3$ divisible by q_2.

Quite generally, Hindu treatises on astronomy and mathematics contain methods of calculation, but do not give proofs. In this respect, they are similar to Greek and Arabic astronomical table sets, whereas in Greek theoretical treatises such as the Elements of Euclid and the Almagest of Ptolemy full proofs are given for every theorem.

I suppose that the Greeks were able to solve Pell's equation, not only for $D = 2$ and $D = 3$, but also for higher values of D, by a systematic application of the Euclidean algorithm. I also suppose that their methods of calculation were copied, without proofs, in Hindu treatises like Brahmagupta's Siddhanta. From these treatises, Jayadeva could learn the method of composition of solutions and of eliminating common factors of the "roots" x and y, but the idea underlying the cyclic procedure, namely the periodicity of the Euclidean algorithm, got lost. The Hindu mathematicians knew how to apply the Euclidean algorithm to *numbers* in order to find their greatest common divisor, but they did not apply it, as far as we know, to incommensurable *line segments*. This explains, I think, why they had to use the pulverizer in their Chakravala process.

If this explanation is accepted, we can draw, hypothetically, a continuous line of development from the Pythagoreans and Theodoros of Kyrene through Archimedes and Eratosthenes to Brahmagupta, Jayadeva, and Bhaskara II.

Part C

Pythagorean Triples

In Chapter 1 we have discussed the construction of Pythagorean triples in Babylonian, Indian, and Chinese sources. We have also mentioned a method used by Diophantos of Alexandria, but we have not discussed his methods in greater detail. I shall now give a summary of the last preserved book Z of the "Arithmetica" of Diophantos, which deals with Pythagorean triples.

Book Z contains 24 problems on right-angled triangles. From the very beginning, Diophantos assumes the sides of the triangles to be integer multiples of a unit of length, and he assumes the integers x,y,z to be of the form

$$(1) \qquad x = 2mn, \quad y = m^2 - n^2, \quad z = m^2 + n^2.$$

For instance, Problem 1 reads:

To find a right-angled triangle such that the hypotenuse minus each of the sides gives a cube.

I have reproduced the translation of Th. Heath: Diophantus of Alexandria (Cambridge 1910, reprinted by Dover, New York 1964). Diophantos himself does not write "the hypotenuse" but ὁ ἐν τῇ ὑποτεινούσῃ, which means something like "the (number of units) in the hypotenuse".

The solution begins with the words:

Let the required triangle be formed from the two numbers s and 3. Now the hypotenuse becomes $s^2 + 9$, one of the rectangular sides $6s$, and the base $s^2 - 9$.

As in Chapter 4, I have denoted the unknown number, for which Diophantos has a special symbol, by s.

The continuation of the solution is very curious:

The hypotenuse less the base is now 18, and this is not a cube. Where does the number 18 come from? It is twice the square of 3. So one has to find (instead of 3) a number such that its square, taken twice, is a cube. Call this numbed s, then $2s^2$ has to be a cube, say s^3, and s becomes 2.

Diophantos always manages to reduce his problems to equations with just one unknown s. If his assumptions don't work, he investigates why they don't work, and replaces one of his assumed numbers (in this case 3) by a new unknown, which he again denotes by the same symbol. In our case, he finds that a possible value for the new unknown is 2. He now replaces his original pair of numbers $(s,3)$ by a new pair $(s,2)$, and he continues:

Now form again a triangle with the aid of the numbers s and 2. The hypotenuse becomes $s^2 + 4$, the height $4s$, and the base $s^2 - 4$. The difference of the hypotenuse and base is a cube.

The other conditions requires that the hypotenuse less the height, that is, $s^2 + 4 - 4s$, is a cube. Now this expression is a square of side $s - 2$. Therefore, if we set $s - 2$ equal to a cube, the work is done.

Take for this cube 8, we get $s = 10$. The triangle is therefore formed from 10 and 2. The hypotenuse is 104, the height 40, and the base 96.

The same pair $(s,2)$, this time with $s < 2$, is used in the solution of Problem 2. The problem reads:

To find a right-angled triangle such that the hypotenuse added to each side gives a cube.

The solution begins thus:

Let us from the required (triangle) from s and 2. The hypotenuse becomes $s^2 + 4$, one of the rectangular sides $4s$, the other $4 - s^2$.

The rest is easy. The sum

$$(s^2+4)+(4-s^2)=8$$

is already a cube. So we have to make the sum

$$(s^2+4)+4s=(s+2)^2$$

equal to a cube. This goal is attained if $s+2$ is a cube between 2 and 4, say

$$s+2=27/8.$$

Therefore $s=11/8$, and the triangle is

$$\left(\frac{135}{64}, \; 5\frac{1}{2}, \; \frac{377}{64}\right)$$

or, if we multiply by 64,

$$(135, 352, 377).$$

Brahmagupta also uses the formula (1) to construct rational right-angled triangles. He presents this general rule in chapter 12, verse 33 of his "Brahma-sphuta-siddhanta". It is pretty certain that Brahmagupta was not influenced by Diophantos. Moreover, as we have seen in Chapter 1, the special case $n=1$ of Brahmagupta's rule (1) was ascribed to Plato by late Greek sources. I suppose that the rule (1) comes from a very early tradition.

In verse 35 of the same chapter, Brahmagupta poses the problem:
Given a side a, to construct a right-angled triangle with rational sides.
Brahmagupta's solution is

$$x = a$$
$$y = 1/2 \left(\frac{a^2}{m} - m\right)$$
$$z = 1/2 \left(\frac{a^2}{m} + m\right).$$

Here m is the difference $z-y$. So, what Brahmagupta does, is to calculate a right-angled triangle of which one side $x=a$ and the difference of the two other sides are given. This is a problem which is also treated in the Babylonian text BM 34568 and in the Chinese "Nine Chapters". We have seen in Chapter 1 that these texts are based on a common pre-Babylonian tradition. Brahmagupta's solution is identical with the Babylonian and Chinese solutions. I suppose it is ultimately derived from the same pre-Babylonian tradition.

Chapter 6

Popular Mathematics

The present chapter will be divided into seven parts:
A. General Character of Popular Mathematics
B. Babylonian, Egyptian and Early Greek Problems
C. Greek Arithmetical Epigrams
D. Mathematical Papyri from Hellenistic Egypt
E. Squaring the Circle and Circling the Square
F. Heron of Alexandria
G. The Mishnat ha-Middot

Part A

General Character of Popular Mathematics

Side by side with the tradition of classical Greek geometry, which is known from the works of Euclid, Archimedes. Apollonios, and Pappos, a more popular tradition existed: a tradition of arithmetical and geometrical problems with numerical solutions, similar to the problems we find in Egyptian, Babylonian, and Chinese collections.

The main difference between the two traditions are:

1°. Classical treatises are based on postulates and axioms and proceed with constructions, theorems and proofs. In the popular tradition, the emphasis is on problems and numerical solutions. Proofs are given only incidentally.

2°. In classical Greek mathematics and in Plato's dialogues, the word Arithmos always means an integer: no fractions are admitted. In popular arithmetic and in the work of Diophantos, fractional solutions are allowed.

3°. In the arithmetical book of Euclid's Elements and also in the work of Diophantos only pure numbers are considered, abstracted from all practical weighing, mixing, and paying problems. In popular arithmetic prices and weights are calculated, apples and nuts are counted.

Plato describes this kind of arithmetic very clearly in book 7 of the "Laws", 819 B-C. The protagonist of the dialogue, the "Athenian" says:

One ought to declare, then, that the freeborn children should learn as much of these subjects as the innumerable crowd of children in Egypt learn along with their letters. First, as regards counting, lessons have been invented for the merest infants to learn, by way of play and fun, modes of dividing up apples and chaplets, so that the same totals are adjusted to larger and smaller groups ... Moreover, by way of play, the teachers mix together bowls made of gold, bronze, silver and the like, and others distribute them, as I said, by groups of a single kind, adapting the rules of elementary arithmetic to play; and thus they are of service to the pupils for their future tasks of drilling, leading and marching armies, or of household management, and they render them both more helpful in every way to themselves and more alert.

It is very remarkable that Plato ascribes this mode of teaching practical arithmetic to the Egyptians.

Part B

Babylonian, Egyptian and Early Greek Problems

Two Babylonian Problems

In the Old-Babylonian text VAT 8389 we find the following problem (see O. Neugebauer: Mathematische Keilschrifttexte I, p. 323):

Per bùr (surface unit) I have harvested 4 gur of grain. From a second bùr I have harvested 3 gur of grain. The yield of the first field was 8,20 more than that of the second. The areas of the two fields were together 30.0. How large were the fields?

For a full understanding of the calculation one has to know that the combined area of the two fields (30,0) is measured in SAR, and the difference in yields in terms of sila, and that

$$1 \text{ bùr} = 30,0 \text{ SAR}$$
$$1 \text{ gur} = 5,0 \text{ sila}.$$

According to Thureau-Dangin, the SAR and the sila are the "scholar's units", in terms of which mathematical calculations are carried out, while bùr and gur are larger, practical units.

The first field yields 4 gur = 20,0 sila per 1 bùr = 30,0 SAR, and the second field 3 gur = 15,0 sila per 30,0 SAR. Call the unknown areas (expressed in SAR) x and y. Then we have to solve two equations with 2 unknowns:

$$\frac{20,0}{30,0} x - \frac{15,0}{30,0} y = 8,20$$
$$x + y = 30,00$$

The Babylonians were fully able to solve the second equation for x, to substitute the result into the first and then to solve for y. Indeed this is what they did in another problem of the same text, in which $x - y$ was given instead of $x + y$. But in the present problem, they followed a different path. They begin by dividing the total area into two equal parts:

Divide 30,0, the sum of the areas, into two parts: 15,0. Thus take 15,0 and again 15,0.

Next they calculate what the yield would be, if each of the fields had an area of 15,0. Every step is worked out in great detail: the reciprocal of 30,0 is multiplied by 20,0 and we find "the wrong yield of grain" 0;40, i.e. the yield of the first field for 1 SAR. Hence the yield of a field of 15,0 SAR would be

$$0,40 \times 15,0 = 10,0.$$

"Keep this in mind", the text says. In the same manner one finds for the second field the "wrong yield of grain" of 0;30 for 1 SAR, and hence a yield of 7,30 for 15,0 SAR.

It is concluded that, if each of the fields had an area of 15,0 SAR, the difference in yield would be

$$10,00 - 7,30 = 2,30.$$

But it is given that the difference is 8,20. "Subtract", the text says:

$$8,20 - 2,30 = 5,50.$$

"Keep 5,50 in mind", says the text and then continues the calculation as follows:

$$0;40 + 0;30 = 1;10.$$

I don't know the reciprocal of 1;10. What must I multiply by 1;10 to obtain 5,50? Take 5,0, since $5,0 \times 1;10 = 5,50$.

Subtract this 5,0 from one of the areas of 15,0 and add it to the other. The first is 20,0, the second 10,0. So 20,0 is the area of the first field, 10,0 that of the other.

A school teacher might explain the procedure to children as follows:

If each of the fields had an area of 15,0 SAR, the difference in yield would be 2,30. It has to be 8,20, so that 5,50 has to be added. For every unit of area, added to the first field and subtracted from the second, the first would produce 0;40 more and the second 0;30 less, so that the difference would be increased each time by $0;40 + 0;30 = 1;10$. This has to be taken 5,0 times to obtain exactly 5,50. Hence the first area must be $15,0 + 5,0 = 20,0$ and the second $15,0 - 5,0 = 10,0$.

In another ancient Babylonian text YBC 4652 (see O. Neugebauer and A. Sachs: Mathematical Cuneiform Texts, New Haven 1945, p. 101) we find a sequence of problems of the following kind:

Problem 7. I found a stone, but I did not weigh it. After I added one-seventh and added one-eleventh, I weighed it: 1 ma-na. What was the original weight of the stone?

The original weight of the stone was 2/3 ma-na, 8 gín, 22 1/2 še.

Note that the gín (shekel) is 1/60 of a ma-na, and the še (barleycorn) 1/180 of a gín. In modern notation the problem is equivalent to the equation

$$(1+1/11)(1+1/7)x=1$$

or

$$\frac{12}{11}\cdot\frac{8}{7}x=1.$$

The solution is, of course,

$$x=\frac{11\times7}{12\times8}=0;48,\ 7,30\ \text{ma-na}$$

$$=48\ \text{gín}\ 22\frac{1}{2}\ \text{še},$$

in accordance with the text.

Egyptian Problems

In the Rhind Papyrus we find a sequence of similar problems, in which an unknown quantity '$ḥ$' (perhaps pronounced as "aha") has to be determined from a linear equation. A typical example is Problem 26:

A quantity whose fourth part is added to it becomes 15.

In modern notation, the equation to be solved is

$$(1+1/4)x=15.$$

The solution is as follows:
Reckon with 4: your are to make their quarter, namely 1.

Total   5.

Reckon with 5 to find 15.

/1   5
/2   10
The result is 3.

Multiply 3 by 4.

1   3
2   6
/4   12
The result is 12.
1   12
1/4   3
Total   15
The quantity is 12,
   its quarter is   3
          Total   15.

Peet, the editor of the Rhind Papyrus, explained this calculation as an application of the method of "false assumption". One starts with an arbitrarily chosen quantity, in our case 4. Four and a fourth part of four give 5. The required result is 15, hence the quantity has to be multiplied by

$$15:5=3.$$

Another explanation was given in the fourth edition of J. Tropfke: Geschichte der Elementarmathematik (Berlin 1980), p. 385. One divides the unknown quantity into 4 equal parts. Five of these parts are equal to 15, so every part is 3, and the original quantity was $4 \times 3 = 12$.

In favour of this interpretation one can adduce the fact that the scribe calculated 4 times 3 and not 3 times 4.

The "Bloom of Thymaridas"

The ancient Pythagorean Thymaridas is mentioned several times by the late Pythagorean Iamblichos, in his "Vita Pythagorica" as well as in his commentary to the Arithmetical Introduction of Nikomachos (p. 11 and 27 and 65 in Pistelli's edition of the commentary).

On p. 11 of the commentary Iamblichos informs us that Thymaridas called the prime numbers "linear" (εὐτυγραμμικοί), which means that their units can be ordered in a straight line, but not in a rectangle.

On p. 62 Iamblichos presents a rule for solving special sets of linear equations, a rule which he calls "Bloom of Thymaridas". In the following discussion I shall base myself on the interpretation given by G. H. L. Nesselmann: Die Algebra der Griechen, p. 232–236.

Iamblichos distinguishes known or determined quantities from the undetermined or unknown (ἀόριστον). He considers an arbitrary number of unknown quantities: let's call them $x_1,\ldots,x_n$. To each of them he adds one additional unknown quantity: let's call it x. The sums $x+x_1,\ldots,x+x_n$ are given: let's call them $a_1,\ldots,a_n$. Also, the sum S of all unknown quantities is given. So we have a set of equations:

$$x+x_1+\ldots+x_n = S$$
$$x+x_1 = a_1$$
$$\cdots$$
$$x+x_n = a_n.$$

It is very remarkable that Iamblichos, probably following an ancient source, formulates the solution in general words, not for special given numbers. His solution is equivalent to the modern formula

$$x = \frac{(a_1+\ldots+a_n)-S}{n-1}.$$

This solution can easily be obtained by a method often employed by Diophantos. As we have seen, Diophantos always manages to reduce his problems to equations in just one unknown s. Now, if one puts $x = s$, the other unknowns will be $a_1 - s, \ldots, a_n - s$, and the sum of all unknowns is

$$(a_1 + \ldots + a_n) - (n-1)s = S.$$

This equation can easily be solved for s. The method just described is equivalent to what we have called the Method of Elimination, which was used by the Babylonians to solve sets of equations in two or three unknowns.

The word Aoriston, which Iamblichos uses to denote an unknown quantity, is the very word used by Diophantos in the expression

$$\pi\lambda\tilde{\eta}\vartheta o\varsigma\ \mu o\nu\acute{\alpha}\delta\omega\nu\ \acute{\alpha}\acute{o}\rho\iota\sigma\tau o\nu,$$

which means an undefined or undetermined number of units. So it seems that the "Bloom of Thymaridas" is just a link in a once continuous chain of traditions from Babylonian texts through Pythagoras and the Pythagoreans to Diophantos.

On p. 63, Iamblichos proceeds to show that other types of equations can be reduced to this, so that the rule does not "leave us in the lurch" in those cases either. Iamblichos wants to determine four numbers such that:

the first and the second are together twice the third and fourth,
the first and the third are together three times the second and fourth,
the first and fourth are together four times the second and third.

Thus we have to solve a set of three homogeneous linear equations

$$x + y = 2(z + u)$$
$$x + z = 3(y + u)$$
$$x + u = 4(y + z).$$

From these equations one obtains

$$x + y + z + u = 3(z + u)$$
$$= 4(y + u)$$
$$= 5(y + z).$$

Since the sum of the four numbers must be divisible by 2, by 3, by 4, and by 5, Iamblichos puts the sum equal to

$$2 \times 3 \times 4 \times 5 = 120$$

Now we have

$$
\begin{aligned}
x+y+z+u &= S &= 120 \\
x+y &= 2(z+u) = (2/3)S = 80 \\
x+z &= 3(y+u) = (3/4)S = 90 \\
x+u &= 4(y+z) = (4/5)S = 96
\end{aligned}
$$

Thus, we can apply the "Bloom of Thymaridas" and find

$$
x = \frac{80+90+96-120}{2} = 73
$$

and
$$
y = 7, \quad z = 17, \quad u = 23.
$$

Iamblichos goes on to apply the same method to the set of equations

$$
\begin{aligned}
x+y &= (3/2)(z+u) \\
x+z &= (4/3)(y+u) \\
x+u &= (5/4)(y+z).
\end{aligned}
$$

Part C

Greek Arithmetical Epigrams

The "Greek Anthology" is a Byzantine collection of epigrams (short poems). In book 14 of this collection (Loeb Classical Library, The Greek Anthology, Vol. 5) we find some forty-six arithmetical problems, most of which are ascribed to a certain Metrodoros. According to R. Keydell ("Der kleine Pauly", Art. Metrodoros) the epigrams are not all due to the same author. The collector of the epigrams must have lived after Diophantos, for in one of the epigrams (No 126) Diophantos is mentioned.

Here a few typical examples:

A. Where are the apples gone, my child? B. Ino has two-sixths and Semele one-eighth, and Autonoe went off with one-fourth, while Agave snatched from my bosom and carried away a fifth. For thee ten apples are left, but I, yes I swear it by dear Cypris, have only this one.

This problem leads to the linear equation

$$
(1-2/6-1/8-1/4-1/5)x = 10+1.
$$

The solution is, of course, $x = 120$.

We have seen that Plato, in the "Laws", already mentions arithmetical problems about apples and nuts. Several problems of this kind are versified in the Greek Anthology.

The following problem also leads to a linear equation in one unknown:

126. This tomb holds Diophantos. Ah, how great a marvel! the tomb tells scientifically the measure of his life. God granted him to be a boy for the sixth part of his life, and adding a twelfth part to this, he clothed his cheeks with down; He lit him the light of wedlock after a seventh part, and five years after his marriage He granted him a son.
Alas! late-born wretched child; after attaining the measure of half his father's life, chill Fate took him. After consoling his grief by this science of numbers for four years he ended his life.

Solution: He was a boy for 14 years, a youth for 7, at 33 he married, at 38 he had a son born to him who died at the age of 42. The father survived him for 4 years, dying at the age of 84.

Other problems lead to two linear equations in two unknowns. Example:

146. A. Give me two minas and I become twice as much as you.
      B. And if I get the same from you I am four times as much as you.

The pair of equations is

$$x+2 \quad =2(y-2)$$
$$4(x-2)=y+2.$$

A very remarkable problem, also from the Anthologia Graeca, has been communicated by G. Wertheim in an Appendix to his translation of the Arithmetica of Diophantos. It reads:

A: I am equal to the second and one-third of the third.
B: I am equal to the third and one-third of the first.
C: And I to one-third of the second and ten mines.

The set of equations is

$$x=y+\tfrac{1}{3}z$$
$$y=z+\tfrac{1}{3}x$$
$$z=\tfrac{1}{3}y+10$$

In Babylonian and Chinese texts we have found quite similar problems leading to sets of linear equations in two or three unknowns. It seems that we have here, in the Greek Anthology, a continuation of a pre-Babylonian tradition.

Part D

Mathematical Papyri from Hellenistic Egypt

In 1972, R.A. Parker published a very remarkable collection of Demotic papyri containing mathematical problems with solutions. His publication is entitled Demotic Mathematical Papyri (Brown University Press,

Providence R. I., and Lund Hymphreys, London 1972). In this publication one finds translations of sixty-five problems. The first forty problems are from a Cairo papyrus written in the first century B.C., the remaining ones from two papyri in the British Museum, both later than the Cairo papyrus, perhaps early Roman. We shall now discuss some of the problems in the papyri.

Calculations with Fractions

Acient Egyptian texts such as the Rhind papyrus use only unit fractions $1/n$ and the fraction $2/3$, for which a special sign existed. The Cairo papyrus also uses mixed fractions m/n. For instance, in Problem 3, the result of the division of 100 by 47 is written as $2 + 6/47$. In this respect, the Cairo papyrus is similar to the Chinese "Nine Chapters" and to early Indian arithmetical textbooks. In the Cairo papyrus even composite fractions such as $11\frac{1}{2}/47$ occur.

Yet the ancient Egyptian tradition of expressing mixed fractions as sums of unit fractions is continued in our Demotic papyri. Thus in the fourth problem of the text BM 10520 (Problem 56 in Parker's publication), the mixed fraction 2/35 is written as a sum of two unit fractions, just as in the Rhind papyrus:

(1) $$1/35 + 1/35 = 1/30 + 1/42.$$

If both sides of (1) are multiplied by 210, one obtains

$$6 + 6 = 7 + 5$$

which is obviously correct. Thus the correctness of (1) can be *verified* afterwards by reducing both sides to the common denominator 210. In the Rhind papyrus the numbers 6, 7, and 5 are written as "auxiliary numbers" below the fractions 1/35, 1/30, and 1/42. This standard method is often applied in the Rhind papyrus in order to verify equalities between fractions.

On the other hand, the scribe of the Demotic papyrus BM 10520 shows the reader how the decomposition (1) can be *obtained*. He calculates in succession:

$$35 = 7 \times 5$$
$$7 + 5 = 12$$
$$12 : 2 = 6$$
$$6 \times 5 = 30$$
$$7 \times 6 = 42.$$

The method is perfectly general. If an odd number n can be decomposed into factors u and v, one has

$$n = uv$$

$$u + v = 2w$$

and

$$1/uv + 1/uv = 1/uw + 1/vw.$$

This identity can be verified by multiplying both sides by uvw. An example is the decomposition

$$1/95 + 1/95 = 1/70 + 1/130$$

which is used in the $(2:n)$ table of the Rhind papyrus. See my paper "The $(2:n)$ Table in the Rhind Papyrus", Centaurus 23, p. 259–274 (1980), and also W. Knorr: Techniques of Fractions in Ancient Egypt and Greece, Historia Mathematica 9, p. 133–171 (1982).

Problems on Pieces of Cloth

Problem 8 of the Cairo papyrus reads in the translation of Parker:
A measure (of cloth) which is 7 cubits (in) height and five cubits (in) width, amounts to 35 cloth-cubits. Take off one cubit from its height, add it to its width. What is that what is added to its width?
The idea is that the total area of the piece of cloth shall be conserved. This implies that the area of what is cut off shall be equal to the area added.
The scribe first calculates the new height as $7 - 1 = 6$. He then states: "The taken-off area makes 5 cloth-cubits." Next he divides 5 by 6, and finnaly he adds the result $5/6$ to the width 5. Problem 9 is similar.

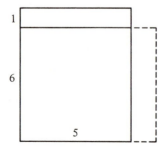

Fig. 68. Cutting off a strip of cloth

Problem 7 also concerns a piece of cloth, but it requires the extraction of a square root. This problem will be discussed later.

Problems on Right-Angled Triangles

Problem 24 of the Cairo papyrus reads:

An erect pole of 10 cubits has its base moved out 6 cubits. Determine the new height and the distance the top of the pole is lowered.

The result is calculated by means of the "Theorem of Pythagoras". The scribe calculates $10 \times 10 = 100$, subtracts $6 \times 6 = 36$, and extracts the square root of 64, which is 8. Subtracting 8 from 10, one obtains 2. Problems 25 and 26 are of the same type.

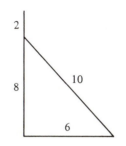

Fig. 69. Right-angled triangle

In problems 27, 28, and 29 the length of the pole and the lowering at the top are given, and one has to find out how far out the foot has been moved.

Problems 30 and 31 are very remarkable because of their similarity with Chinese and Babylonian problems. Problem 30 reads: Given 2 cubits as the distance the top of an erect pole is lowered when its base is moved out 6 cubits, determine its original height. If the sides of the right-angled triangle are called x, y, z, we are given $x = 6$ and $z - y = 2$. This is what I have called in Chapter 2 a problem of Type 4. The solution is the same as in the Chinese and Babylonian texts.

R. A. Parker, who published and translated the Cairo papyrus, concludes that a Babylonian influence is "only too likely and can in certain cases be documented". He notes the close similarity of problems 24 to 31 to Babylonian problems. To this he adds (Demotic Mathematical Papyri, p. 6):

Elsewhere I have shown that the content of a demotic papyrus of the Roman period concerned with celestial omina can be definitely ascribed to Babylonia, the transmission of such literature having taken place during the Persian rule of Egypt, the late sixth and fifth centuries B.C.[4] It is very likely, indeed, that some amount of Babylonian mathematical literature also came to Egypt at the same time, and problems 24 to 31, with their use of the "Pythagorean" theorem, known centuries earlier in Babylonia, appear to be direct confirmation of such transmission.

Parker's footnote 4 reads:

Richard A. Parker, A Vienna Demotic Papyrus on Eclipse- and Lunar-Omina, Brown Egyptological Studies, vol. 2 (Providence 1959), pp. 28–34.

Approximation of Square Roots

Problem 62 (in BM 10520) reads:

Cause that 10 reduce to its square root.

The problem is, to extract the square root of 10. The problem is solved by an approximation thus:

You shall reckon 3, 3 times: result 9, remainder 1.
1/2 (of 1): result 1/2.
You shall cause that 1/2 make part of 3: result 1/6.
You shall add 1/6 to 3: result $3+1/6$. It is the square root.

As a check, the square of $3+1/6$ is calculated. It is $10+1/36$.
The approximation used here is

$$(2) \qquad \sqrt{a^2+b} \sim a+b/2a.$$

The same method of approximating square roots was also known to the Babylonians and to Heron of Alexandria, as we shall see later.

Problem 7 of the Cairo text reads:

If it is said to you: Have sailcloth made for the ships,
and it is said to you: Give 1000 cloth-cubits (i.e. square cubits) to one sail,
have the height of the sail be (in the ratio) 1 to 1 1/2 the width,
(here is) the way of doing it.

In the course of the calculation, the scribe has to extract the square root of 1500. He obtains

$$(3) \qquad 38+2/3+1/20.$$

As Parker notes in his commentary, this result cannot be obtained by the formula (2). This formula always yields results larger than the true value of the square root, whereas the approximation (3) is less than the true value.

I think we can explain the approximation (3) by assuming a Babylonian origin of the method of solution. In the sexagesimal system we have

$$\sqrt{1500} = 38;43,47,\ldots$$

which may be rounded off to

$$38;43 = 38+2/3+1/20.$$

Two More Problems of Babylonian Type

Problems 34 and 35 of the Cairo papyrus deal with rectangular plots of land, of which the area and the diagonal are given. Problem 34 reads:

19. A rectangular plot of land of 60 square cubits, the diagonal
20. being 13 cubits. Now how many cubits does it make to a side? You shall reckon 13 times 13: result 169.

21. You shall reckon 60 twice: result 120. You shall add it to 169: re-
    sult 289.
22. Cause that it reduce to its square root: result 17. You shall take
    the excess of 169
23. against 120: result 49. Cause that it reduce to its square root: re-
    sult 7. Subtract it from 17:
24. remainder 10. You shall take to it 1/2: result 5. It is the width.
25. Subtract 5 from 17: remainder 12. It is the height.
26. You shall say: "Now the plot of land is 12 cubits by 5 cubits."
27. To cause that you know it. Look. You shall reckon 12 times 12:
    result 144.
28. You shall reckon 5 times 5: result 25. Result 169. Cause that it re-
    duce to its square root: result 13.
29. It is the diagonal of the plot.

In the last three lines the scribe shows that the solution found in line 26
is correct. The check is based on the Theorem of Pythagoras: if the sides
are x and y, the diagonal d is calculated as

$$d = \sqrt{x^2 + y^2}\,.$$

This theorem being known, the problem can be reduced to a pair of
equations

(4)
$$x^2 + y^2 = d^2 = 169$$
$$xy = A = 60\,.$$

From these two equations one deduces, in accordance with the text,

$$(x+y)^2 = d^2 + 2A = 289$$
$$(x-y)^2 = d^2 - 2A = 49\,,$$

hence

$$x + y = 17$$
$$x - y = 7$$
$$y = \tfrac{1}{2}(17 - 7) = 5$$
$$x = \tfrac{1}{2}(17 + 7) = 12\,.$$

In Problem 12 of the ancient Babylonian text BM 13901 (O. Neugebauer, Mathematische Keilschrifttexte III, p. 7) a pair of equations of the
same kind is solved, namely

(5)
$$x^2 + y^2 = 21,40$$
$$xy = 10, 0\,.$$

The Babylonian solution is unnecessarily complicated. The scribe first
computes, by a Babylonian standard method, the squares x^2 and y^2 from

the pair of equations

$$x^2 + y^2 = 21,40$$
$$x^2 y^2 = (10, 0)^2$$

and next he finds x and y by extracting square roots. The Egyptian solution is simpler. It is based on the identities

$$(x+y)^2 = x^2 + y^2 + 2xy$$
$$(x-y)^2 = x^2 + y^2 - 2xy.$$

Part E

Squaring the Circle and Circling the Square

An Ancient Egyptian Rule for Squaring the Circle

The ancient Egyptians calculated the area A of a circle of diameter d as

$$A = (8/9\,d)^2.$$

This rule is applied in problem 50 of the Rhind mathematical papyrus (see: R.J. Gillings: Mathematics in the Time of the Pharaos, p. 139). In the enunciation of this problem, the diameter of the circle is supposed to be 9 *khet*. A *khet* is 100 royal cubits, and a square *khet* is called a *setat*. The calculation goes as follows ($\bar{9}$ meaning 1/9):

Take away $\bar{9}$ of the diameter, namely, 1.
The remainder is 8.
Multiply 8 by 8.
It makes 64.
Therefore it contains 64 setat of land.

Do it thus.

1	9
$\bar{9}$	1

The remainder is 8.

1	8
2	16
4	32
\8	\64

The area is 64 setat.

For the sake of convencience, every estimate of π will also be denoted by π. Thus, the ancient Egyptian estimate can be written as

$$\pi = \left(\frac{16}{9}\right)^2 = 3.1605\ldots.$$

If this approximation is adopted, the problems of squaring the circle and circling the square are easy to solve. If a circle is given, one draws a square with side $(8/9)d$. Conversely, if a square is given, one adds one eighth to its side, and one obtains the diameter of a circle of nearly the same area.

Problem 48 of the Rhind papyrus is based on the same approximation. The scribe compares the area of a circle of diameter 9 with the area of a square having side 9 as follows:

The circle of diameter 9		The square of side 9	
1	8 setat	\1	9 setat
2	16 setat	2	18 setat
4	32 setat	4	36 setat
\8	64 setat	\8	72 setat
		Total	81 setat

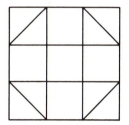

Fig. 70. Octagon inscribed in a square

The drawing accompanying the text shows an octagon inscribed in a square. In his book "Vorgriechische Mathematik I", K. Vogel has given a plausible explanation of the drawing and the calculation. If a square is divided into 9 equal squares, and if triangles at the four vertices are cut off as indicated in Fig. 70, the remaining octagon has an area of 7/9 of the original square, and this is a reasonable approximation of the area of the circle. If one wants to find a square equal in area to the octagon, one has to extract the square root of 7/9. If this square root is approximated by the well-known formula

$$\sqrt{a^2 - b} \sim a - b/2a$$

one obtains

$$\sqrt{1-2/9} \sim 1-1/9 = 8/9,$$

which is just the solution given in the Rhind papyrus.

Circling the Square as a Ritual Problem

In the Śulvasutras the problem of circling the square takes the form of a ritual problem. One has a square altar, and one wishes to construct a circular altar having the same area. The solution, as given in the Śulvasutras[15], is as follows:

In square $ABCD$, let M be the intersection of the diagonals. Draw the circle with M as center and MA as radius; and let ME be a radius perpendicular to AD and cutting AD in G. Let $GN = 1/3\ GE$. Then MN is the radius of a circle having area equal to the square $ABCD$.

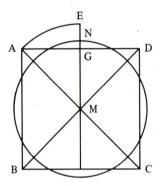

Fig. 71 Circling the square

In modern notation, this construction amounts to putting

$$\pi \left(1 + \frac{\sqrt{2}-1}{3}\right)^2 = 4$$

or

$$\pi = \left(\frac{6}{2+\sqrt{2}}\right)^2 = 3.088\ldots.$$

An Egyptian Problem

Problem 32 of the Cairo papyrus (first century B.C.) reads in the translation of Parker:

15 A. Seidenberg: The Ritual Origin of Geometry, Archive for History of the Exact Sciences 1, p. 488–527, especially p. 515.

(As for) a piece (of land) that amounts to 100 square cubits that is square, if
    it is said to you: Cause that it make a piece (of land)
that amounts to 100 square cubits that is round, what is the diameter? Here
    is its plan.
You shall add the 1/3 of 100 to it: result: 133 1/3. Cause that it reduce to
    its square root: result 11 1/2 1/20.
You shall say: "11 1/2 1/20 is the diameter of the piece (of land) that
    amounts to 100 square cubits."

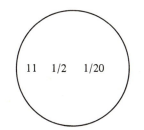

Fig. 72. Drawing to problem 32 of the Cairo papyrus

Commentary: A square piece of land having an area of 100 square cubits (and hence a side of 10 cubits) is given, and it is required to find a circular piece of land having the same area. This problem of "circling the
square" is the inversion of the famous Greek problem of "squaring the circle".

If the area is A and the diameter of the circle d, the scribe calculates d
according to the formula

$$d=\sqrt{A+(1/3)A}$$

which is equivalent to

$$d^2=(4/3)A$$

or to

(6) $$A=(3/4)d^2.$$

In the next line (not reproduced here) the scribe calculates the circumference of the circle as

(7) $$C=3d.$$

The Babylonians too used to compute area and circumference of the
circle by means of (6) and (7). We can express the same facts by saying that
the Babylonians and the Egyptian author of Problem 32 assumed $\pi=3$.

In the Cairo text the square root of $133 + 1/3$ is approximated by

(8) $11 + 1/2 + 1/20$.

This approximation cannot be obtained by the formula

(9) $\sqrt{a^2 + b} = a + b/2a$,

but if one calculates the square root of $133;20$ in the sexagesimal system, one obtains

$$11;32,49$$

which can be rounded up to

$$11;33 = 11 + 1/2 + 1/20 .$$

Area of the Circumscribed Circle of a Triangle

Problems 36 and 37 of the Cairo papyrus are very remarkable. Problem 36 reads:

   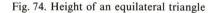
6

Fig. 73. Equilateral triangle inscribed in a circle   Fig. 74. Height of an equilateral triangle

A plot of land – if a large triangle (is) its middle, which is 12 divine-cubits
... segments, what is the area of the plot of land? Here is its plan.
You shall say: "Now it (the plot) makes 4 pieces of land. Look.
They are 1 bare triangle and 3 triangle segments also."
... Here is the (area) of the bare triangle.

You shall reckon 12 times 12: result 144.
You shall reckon 6 times 6: result 36.
Subtract from 144: remainder 108.
Cause that it reduce to its square root:
     result 10 1/3 1/20 1/120.
It is the middle height of the bare triangle.
The number of (the) base (is) 12 cubits.
     Look. Their half (is) 6. Its middle height (is) 10 1/3 1/20 1/120. You shall reckon it 6
     times: result 62 1/3 1/60 cubits. It is the area (of the triangle).

The triangle is supposed to be equilateral. Its height h is correctly calculated by means of the Theorem of Pythagoras:

$$h = \sqrt{12^2 - 6^2} = \sqrt{108}.$$

The approximation to the square root of 108 used in the text:

(10) $h = 10 + 1/3 + 1/20 + 1/120 = 10;23,30$

is excellent, for the correct value, computed in the sexagesimal system, is

$$\sqrt{108} = 10;23,32,\ldots.$$

Multiplying the height of the triangle with half of its base, the scribe obtains the area of the triangle. In the sexagesimal system, the calculation would be very simple:

$$T = 6 \times 10;23,30 = 62;21.$$

The scribe writes this correctly as

$$T = 62 + 1/3 + 1/60.$$

Next he multiplies the height by 1/3. The result

$$3 + 1/3 + 1/10 + 1/60 + 1/120 + 1/180 = 3;27,50$$

is correctly interpreted as the arrow a of one of the three circle segments (see Fig. 75, which is a transcription of a drawing in the Cairo papyrus).

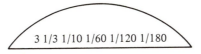

3 1/3 1/10 1/60 1/120 1/180

Fig. 75. Arrow of a circle segment

Next the area of the circle segment is computed as

(11) $F = \frac{1}{2}(s + a)a$

where $s = 12$ is the base of the segment. In Chapter 2 we have seen that this formula also occurs in Heron's "Metrica" and in the Chinese "Nine Chapters". The result of the multiplication (11) is written as

(12) $F = 26 + 5/6 + 1/10 \quad (= 26;56),$

which is not correct. The correct result would be $F = 26;47$.

The result (12) is now multiplied by 3 and added to the area T of the triangle. The result 143;9 is correctly written as

$$T+3F=143+1/10+1/20 \text{ square cubits.}$$

"It is the area of the entire plot of land", says the scribe.

Next the circumference C of the circle is calculated as 3 times the diameter, and one quarter of the circumference is multiplied by the diameter in order to obtain another estimate of the area:

$$(1/4)C \cdot d = 143+5/6+1/10+1/30 \ (=143;58).$$

The scribe now compares this estimate with his earlier result $T+3F$ and finds a difference of

$$2/3+1/10+1/120 \ (=0;49).$$

He says: "This is the error of squaring."

I feel we may be pretty sure that the whole calculation was originally performed in the Babylonian sexagesimal system, and that the resulting sexagesimal fractions were afterwards converted into sums of fractions.

Area of the Circumscribed Circle of a Square

Problem 37 of the Cairo papyrus is similar to problem 36. In a circle of diameter 30 cubits a square is inscribed (see Fig. 76). The side of the square is calculated thus:

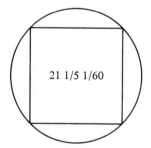

21 1/5 1/60

Fig. 76. Square inscribed in a circle

You shall reckon 30 times 30: result 900. You shall take to it half: result 450. Cause that it reduce to its square root: result 21 1/5 1/60.
They are the measurements of the piece. To cause that you know it.

The square root of 450, calculated in the sexagesimal system, would be

$$21;12,47,\dots$$

The text has

$$s = 21 + 1/5 + 1/60 = 21;13.$$

This number is multiplied by itself. The result is, if minute fractions are neglected, 450.

Next the areas of the four segments are computed. Subtracting the side s from the diagonal, one obtains

$$30 - 21;13 = 8;47 = 8 + 2/3 + 1/10 + 1/60$$

as the scribe correctly states. Half of this is the arrow of the segment

$$a = 4 + 1/3 + 1/20 + 1/120$$

In the papyrus this number is written inside the segment (see Fig. 77). Next the area of the segment is calculated by means of the rule (11):

$$F = \tfrac{1}{2}(s+a)a.$$

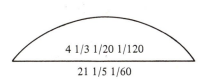

4 1/3 1/20 1/120

21 1/5 1/60

Fig. 77. Arrow of a circle segment

The result $F = 56 + 1/4$ is multiplied by 4 and added to 450, the area of the square. Result 675.

From the problems 36 and 37 one gets the impression that the rule (11) for computing the area of a circle segment was meant as an auxiliary rule. The main problem was: *to evaluate the area of the circle.* For this purpose the inventor of the method inscribed a triangle or a square, computed its area, and evaluated the areas of the surrounding segments by means of the approximation (11).

Three Problems Concerning the Circle Segment in a Babylonian Text

One of the most extensive Babylonian mathematical texts is BM 85 194, published in Vol. IX of the series "Cuneiform Texts in the British Museum", No 8–13. According to O. Neugebauer (Mathematische Keilschrifttexte I, p. 142) the text was written shortly before or after the end of the reign of the dynasty of Hammurabi.

In § 12 of this text (Reverse I, lines 33–43) two problems concerning the chord s and the arrow a of a circle segment are solved. I shall retain Neugebauer's notations (see Fig. 78):

$$a = \text{arrow}$$
$$s = \text{chord}$$
$$d = \text{diameter}$$
$$F = \text{area.}$$

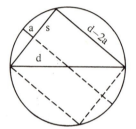

Fig. 78. Arrow a and chord s of a circle segment

In the first problem, the circumference $C = 60$ and the arrow $a = 2$ are given. The diameter d is computed as

$$d = C/3 = 20,$$

and the chord s is calculated by the correct formula

(13) $$s = \sqrt{d^2 - (d - 2a)^2} = 12.$$

In the second problem, $C = 60$ and $s = 12$ are given, and it is asked to find the arrow a. Once more, d is put equal to $C/3$. The correct formula for a would be

(14) $$a = \tfrac{1}{2}(d - \sqrt{d^2 - s^2}).$$

The scribe correctly computes $d^2 - s^2 = 4,16 = 256$. The square root would be 16. The scribe takes 16, without saying that it is the square root of 4,16. He now calculates

$$\sqrt{16} \quad = 4$$
$$(1/2) \cdot 4 = 2,$$

and he says that 2 is the required arrow. The result happens to be correct, but the method of calculation is nonsense. Obviously, the scribe knew beforehand what result had to come out, but he did not understand the method of calculation. Perhaps he was copying a more ancient tablet from which a part was broken off or not intelligible to him.

In § 17 of the same text (Reverse III, lines 1–6) the problem is, to determine the area F of a circle segment, when the arc $b=60$ and the chord $s=50$ are given. The solution given in the text is unintelligible to Neugebauer as well as to me. The scribe first multiplies the difference

$$b-s=10$$

with the chord s, thus obtaining

$$s(b-s)=8,20.$$

From this he subtracts the square of a quantity $RI=10$ in order to obtain the area F of the segment. The square of 10 is 1,40, but the subtraction

$$F=8,20-1,40=7,30$$

is quite wrong. Our earlier conclusion that the scribe was not very intelligent is confirmed by this false subtraction.

However this may be, it is clear that our three Babylonian problems are closely related to the problems 36 and 37 of the Cairo papyrus. The wording of the problems and the method of solution are similar, and so are the drawings. In all cases the object of investigation is a circle segment, and one of the main tools is the Theorem of Pythagoras. The circumference of the circle is always assumed to be 3 times the diameter. The fractions used in the Cairo papyrus are all finite sexagesimal fractions, converted into sums of common fractions.

Shen Kua on the Arc of a Circle Segment

In the year A.D. 1086 the Chinese scientist *Shen Kua* wrote a book *Mêng Chhi Pi Than* (Dream Pool Essays). According to Needham (Science and Civ. in China, Vol. 3, p. 38) this book is not a formal mathematical treatise. It contains notes on almost every science known in his time. but there is much of algebraic and geometric interest to be found in it.

In his Chapter 18, Shen Kua explains a method for computing the arc bounding a circle segment. In Fig. 79 I have used the same letters as before to denote line segments, namely

$$d = \text{diameter}$$
$$s = \text{chord}$$
$$a = \text{arrow}$$
$$b = \text{arc}.$$

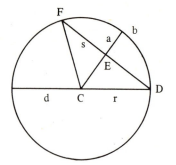

Fig. 79. Arrow, arc and chord of a circle segment

If d and a are given, Shen Kua instructs us to compute the radius $r=CD$ by halving the diameter d, next to compute $CE=r-a$, and then $DE=\frac{1}{2}s$ as

(15) $$\tfrac{1}{2}s=\sqrt{r^2-(r-a)^2}\,.$$

Now the arc b is calculated as

(16) $$b=s+2a^2/d\,.$$

To explain formula (16), let us consider the circle sector bounded by the arc b and the two radii CD and CF drawn towards its end points. The sector is the sum of a triangle and a circle segment. The area of the triangle is

(17) $$T=\tfrac{1}{2}s(r-a),$$

and the area of the segment may be computed by the ancient formula (11):

(18) $$F=\tfrac{1}{2}(s+a)a\,.$$

Adding the two expressions for T and F, one obtains for the area of the sector the expression

$$A=T+F=\tfrac{1}{2}sr+\tfrac{1}{2}a^2\,.$$

Now the area of the sector is known to be

(19) $$A=\tfrac{1}{2}br\,.$$

Equating the two expressions for A, one obtains the equation

$$\tfrac{1}{2}br=\tfrac{1}{2}sr+\tfrac{1}{2}a^2$$

which can be solved for b. The result is just the formula (16).

This text clearly belongs to the same tradition as the Egyptian and Babylonian texts discussed earlier. The formulae (17), (18), and (19), from which we have derived the final result (16), all occur in the Egyptian and Babylonian texts. We are bound to assume a continuous tradition during more than 2700 years from the end of the Hammurabi dynasty to the time of Shen Kua.

From the Cairo papyrus I have already concluded that the rule for the area of a circle segment was only an auxiliary rule, the main aim being the calculation of the area and circumference of a circle. This conclusion is confirmed by the Chinese text. Here the main aim is clear: to calculate the arcs into which a circumference can be divided. Shen Kua himself uses the expression "division of a circumference".

As usual, The Chinese text is better than the Babylonian and Egyptian texts. The latter texts were written by scribes, sometimes even by ignorant scribes, whereas the Chinese text is clearly due to a competent mathematician. It presents general rules instead of special calculations. Also the rule (15) is motivated as an application of the "Theorem of Pythagoras", whereas the same rule is used in the Babylonian and Egyptian texts without any motivation.

It seems to me that these facts confirm my earlier thesis (see Chapters 1 and 2), namely:

The logical structure of pre-Babylonian mathematics is most faithfully preserved in our Chinese mathematical texts.

Part F

Heron of Alexandria

Heron, who lived at Alexandria in the first century A.D., was a very prolific and popular author. In his works on applied mechanics and physics he describes all sorts of mechanical instruments, some of which he seems to have invented himself. The titles are:

> On the Dioptra
> Pneumatica
> On the Construction of Automata
> On the Construction of Engines of War
> Cheirobalistra (on Catapults)
> Katoptrika (on Mirrors)
> Mechanics (extant in Arabic).

In Heron's Opera Omnia we also find several geometrical treatises under titles like Geometrica, Stereometrica, Geodaesia, and Mensurae, containing Heronic material with later additions. It seems that only two of the extant geometrical treatises are genuine, namely

> Metrika
> On Definitions.

In addition to these, several fragments of Heron's commentary on Euclid's Elements have been preserved, some by Proklos and most of them by An-Nairizi in his Arabic commentary to Euclid's Elements. See for these fragments:

Proklos: A Commentary on the First Book of Euclid's Elements, translated by G. R. Morrow, Princeton 1970, p. 154, 249, 252 and 270.

R. O. Besthorn and J. L. Heiberg: Codex Leidensis 399, I: Euclidis Elementa ex interpretatione Al-Hadschadschii cum commentariis Al-Narizii, Arabic and Latin, Copenhagen 1893–1911.

The Date of Heron

The long-debated question of Heron's date has been settled by O. Neugebauer[16], who pointed out that Heron, in his "Dioptra", describes a lunar eclipse observed at Alexandria ten days before the vernal equinox and beginning in the fifth hour of the night. Neugebauer's calculations show clearly that this can only be the eclipse of A. D. 62, March 13. Most probably, Heron observed the eclipse himself or used a contemporary record.

Heron's Commentary to Euclid

According to Proklos, Heron reduced the number of axioms to three. He also gave alternate proofs of I 19, I 20, and I 25.

More important for our purpose is a purely algebraic method to prove the propositions II 2–10 of Euclid's Elements. These propositions belong to what we have called Euclid's "geometrical algebra". Heron shows that these propositions can be proved "without figures" as consequences of II 1.

As we have seen in Chapter 3, Euclid's proposition II 1 corresponds to the algebraic formula

$$a(b+c+\ldots)=ab+ac+\ldots.$$

Heron explains that it is not possible to prove II 1 without actually drawing the rectangles $ab, ac,$ etc., but that the following propositions up to II 10 can be proved by merely drawing one line. In fact, II 2 and II 3 are special cases of II 1. The next proposition II 4 corresponds to

$$(a+b)^2=a^2+b^2+2ab.$$

---

16 O. Neugebauer: Über eine Methode zur Distanzbestimmung Alexandria–Rom bei Heron. Kgl. Danske Vidensk. selsk., Hist.-fil. Meddel. 26, No. 2 and No. 7 (1938 and 1939).

It can be proved by a repeated application of II 1, as follows:

$$(a+b)(a+b)=(a+b)a+(a+b)b$$
$$=a^2+ba+ab+b^2$$
$$=a^2+b^2+2ab.$$

To translate this algebraic proof into the Greek terminology, one has to draw one line segment ABC only:

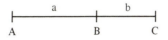

Fig. 80. The segment $a+b$

In a similar way, Heron proves the remaining propositions II 5–10.

Heron's Metrika

This very popular treatise consists mainly of prescriptions for computing areas and volumes. Whereas Euclid operates with the line segments, polygons, circles, and solids themselves, without ever using words like "length", "area", or "volume", Heron is mainly concerned with the numerical values of areas and volumes. When Heron meets an irrational square root, he approximates it by a rational number. He freely uses the results of the great Greek geometers, notably of Archimedes, and he sometimes even gives proofs, but his aim is always: to *calculate* something. In this respect, he continues the tradition of the Babylonians and Egyptians.

Heron first computes the area of a rectangle by multiplying the lengths of two perpendicular sides. His proof is valid only if the two lengths are integers, but afterwards he applies the same rule even to irrational sides. In one place he says: If AB and BC are the sides of a rectangle, the area will be the square root of $AB^2 \cdot BC^2$. Geometrically, one cannot form a product of two squares: Euclid and Apollonios never do such a thing. Like all engineers and physicists, Heron supposes every line segment to have a definite numerical length.

Heron next calculates the areas of triangles, trapezia, rhombs, and other quadrangles. For the area of a triangle with given sides a, b, c he presents the well-known formula

(1)
$$T=\sqrt{s(s-a)(s-b)(s-c)}$$
$$s = \tfrac{1}{2}(a+b+c)$$

with a very nice geometrical proof. According to Al-Biruni[17], the formula
(1) is due to Archimedes. As an example, Heron takes a triangle with sides
7, 8, and 9. He finds

$$T=\sqrt{720}$$

and he calculates an approximation by means of the formula

(2) $\sqrt{a^2\pm b}=a\pm b/2a.$

This formula is not found in any one of the extant classical Greek treatises, but it was known to the Babylonians and Egyptians, as we have seen
in Part D of the present chapter.

Very interesting are the sections in which Heron calculates the areas of
regular polygons with 5 or 6 or ... up to 12 sides. Some of the calculations
are based on exact geometrical theorems, others are only approximations.
In each case, square roots are approximated as before.

Circles and Circle Segments

Quoting Archimedes, Heron calculates the area of a circle as $3+1/7$
times the square of the radius, and the circumference as $3+1/7$ times the
diameter.

Heron next considers a circle segment less than a semicircle. He reports
that "the ancients" calculated the area of the segment as

(3) $F=\frac{1}{2}(s+a)a$

where s is the chord and a the arrow. We have seen that the same approximation was also used in the Cairo papyrus and in the Chinese "Nine
Chapters". Heron thinks that "the ancients", who used this inaccurate
method, followed those who assumed the circumference of a circle to be
three times the diameter. For if this is assumed, the area of a semicircle will
be found in accordance with (3).

But, says Heron, "those who investigated the matter more accurately"
added to the right side of (3) a correction term, thus obtaining a better approximation

(4) $F=(1/2)(s+a)a+(1/14)(\frac{1}{2}s)^2.$

---

17 Al-Biruni: Das Buch über die Sehnen (The Book Concerning the Chords). Bibliotheca
   Mathematica XI 3, p. 11–78.

If one applies this formula to a semi-circle of radius r, one obtains

$$F = 3/2\, r^2 + 1/14\, r^2$$

and hence

$$2F = (3 + 1/7)\, r^2.$$

This, Heron says, is in accordance with the other method which assumes the circumference to be $3 + 1/7$ times the diameter. We know that the latter estimate is due to Archimedes, so "those who examined the matter more accurately" must have lived between Archimedes and Heron.

Heron next says that the approximation (4) is appropriate for segments in which the base does not exceed 3 times the height. This assertion is correct, for as long as s lies between 2 h (semi-circle) and 3 h, the error of the approximation (4) remains less than $1/300$ of the true value.

For the case $s > 3$ h, Heron presents another, very good approximation. He proves that every circle segment is larger than $4/3$ times the triangle having the same base and height. His proof is correct and very much similar to a proof given by Archimedes in his "Quadrature of the Parabola". As we have seen in Chapter 3, Archimedes approximated a parabola segment by a sum of terms like

$$A + B + \Gamma + \Delta + E$$

in which each term is just $1/4$ of the preceding term. The sum of this series tends to $4/3\, A$ if the number of terms is increased. On the other hand, Heron approximates his circle segment by a sum of terms in which each term is slightly larger than $1/4$ of the preceding term. Thus, if one takes a sufficiently large number of terms, the sum will be larger than $(4/3)\, A$.

At the end of his discussion Heron concludes that by adding $1/3$ to the area of the triangle, one obtains very nearly the area of the circle segments. Whoever invented this very accurate method of approximation and this beautiful proof must have been an excellent mathematician.

Who was it? Only one great mathematician living between Archimedes and Heron is known, namely Apollonios of Perge. He is a most likely candidate.

Apollonios' "Rapid Method"

It is known that Apollonios wrote a treatise entitled "Rapid Delivery" (Okytokion), in which he improved on Archimedes' estimate of π. Archimedes had used inscribed and circumscribed polygons of 96 sides. To improve on his estimate $3 + 1/7$, one might increase the number of sides and calculate the perimeters or areas of the polygons more accurately, but this would not be a "rapid" method. Another possibility would be to calculate the area of an inscribed polygon having a moderate number of sides and to add an estimate of the sum of the remaining circle segments. If Apollonios used this method, he would need an approximation for the area of a circle segment. Using the estimates explained in Heron's "Metrika", he would really have a rapid and accurate method. For this reason, I guess that Heron's approximations are due to Apollonios.

Paul Tannery already conjectured (Recherches sur l'histoire de l'astronomie ancienne, Paris 1893, p. 64–66) that Apollonios' estimate of π was

$$\pi = 3.1416.$$

As we shall see in Chapter 7, this estimate was known to Liu Hui (3rd century A. D.), to Aryabhata (6th century), and to Bhaskaracarya (12th century). I suppose they all had it from Apollonios. Arguments in favour of this hypothesis will be presented in Chapter 7.

Volumes of Solids

Heron finds the volume of a rectangular block by multiplying length, breadth, and height. More generally, the volume of a prism or cylinder is found by multipliying the base area with the height.

Heron also knows that the volume of a cone or pyramid is one third of the volume of a cylinder or prism on the same base having the same height.

A truncated pyramid is measured as the difference of two pyramids. Just so, a truncated cone is a difference of two cones. More complicated polyhedra are divided into prisms and pyramids.

Quoting Archimedes, Heron determines the volume of a sphere as 2/3 of the volume of a circumscribed cylinder. For a segment of a sphere, he likewise uses Archimedes' result.

Next, Heron calculates the volume T of a torus. Let C be a circle in the plane of the line AB (Fig. 81), and let the torus be generated by the rotation of the circle about the line AB. Let r be the radius of the circle, and b the distance of its centre from the axis AB. Let the diameter EF be parallel to AB. Heron cites Dionysodoros, who proved that the torus is to a cylinder

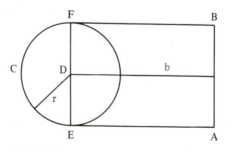

Fig. 81. A rotating circle generates a torus

with height $AB=2r$ and base radius b as the circle C is to half the rectangle $ABEF$. In modern notation, we have the proportion

$$T:\pi b^2 \cdot 2r = \pi r^2 : rb$$

from which T may be computed. Once more, Heron uses the estimate $\pi = 22/7$.

Heron also presents another method which yields the same result. He says: a circle with centre A and radius $b = 14$ will have diameter 28 and circumference 88. Now if the torus is straightened, it becomes a cylinder with height 88. The base of the cylinder is a circle with diameter $2r = 12$, so the volume of the cylinder is known. The result is the same as before. This derivation is a typical piece of popular mathematics: the method is plausible, but a proof is not given.

Making use of Archimedes' treatise "Ephodos", Heron next calculates the volumes of two remarkable solids. The first is cut off from a cylinder by an oblique plane, and the other is an intersection of two cylinders inscribed in a cube. The same intersection was also considered by Chinese authors, as we shall see in Chapter 7.

At the end of his chapter on volumes, Heron calculates the volumes of the five regular solids.

I shall leave aside the purely geometrical part of Heron's next chapter "On Division of Figures", and restrict myself to his method of calculating cube roots.

Approximation of a Cube Root

Heron wants to calculate the cube root of 100, and says:

Take the nearest cube numbers to 100 both above and below; these are 125 and 64.
Then $125 - 100 = 25$,
and $100 - 64 = 36$.
Multiply 5 into 36; this gives 180. Add (4 times 25, or) 100, making 280. Divide 180 by 280; this gives 9/14. Add this to the side of the smaller cube: this gives $4 + 9/14$. This is as nearly as possible the cube root ("cubic side") of 100 units.

This translation is taken from Th. Heath: A History of Greek Mathematics, p. 341. Heath gives a clear account of the explanations of Heron's method given by Wertheim and Eneström. If N is the number from which we have to extract the cube root, Heron first instructs us to find two integers a and $a+1$ so that

$$a^3 < N < (a+1)^3.$$

Next, Heron computes

$$N - a^3 = d_1 \quad \text{and} \quad (a+1)^3 - N = d_2.$$

A linear interpolation of the function x^3 between a and $a+1$ would yield the approximation

$$a + \frac{d_1}{d_1 + d_2}.$$

Instead of this, Heron used the formula

(5) $$\sqrt[3]{N} \sim a + \frac{(a+1)d_1}{(a+1)d_1 + a d_2}.$$

Eneström has explained how this formula can be justified by a plausible approximation, neglecting cubes of small numbers.

Another explanation would be as follows. The function $x^3 - N$, which we want to make zero, has a rapidly increasing slope, so we cannot well approximate it by a linear function. However, if we divide it by x, we obtain a function

$$\frac{x^3 - N}{x} = x^2 - \frac{N}{x},$$

which is more nearly linear. The values of the function at $x = a$ and $x = a + 1$ are

$$-d_1/a \quad \text{and} \quad +d_2/(a+1).$$

Interpolating linearly between these values, one obtains

(6) $$\sqrt[3]{N} \sim a + \frac{d_1/a}{d_1/a + d_2/(a+1)}$$

which is equivalent to (5).

Part G

The Mishnat ha-Middot

The Mishnat ha-middot is a Hebrew treatise on mensuration. It was edited with an English translation and an excellent commentary by Solomon Gandz in Quellen und Studien zur Geschichte der Mathematik, Series A: Quellen (Springer, Berlin 1932).

According to Gandz, the author of the treatise was most probably Rabbi Nehemiah, who lived about A. D. 150. To me, Gandz's arguments in favour of the ascription to Nehemiah are convincing, but see G. B. Sarfatti: Mathematical Terminology in Hebrew Scientific Literature of the Middle Ages (Jerusalem, Magnes 1968), p. 58–60.

As we shall see, much of the material in the Mishnat ha-Middot comes ultimately from the "Metrika" of Heron. However, Heron sometimes gives proofs and quotes Archimedes, whereas the Mishnat presents only rules of calculation without proofs. In this respect, the Mishnat is similar to the pseudo-Heronic "Geometrika".

The treatise first deals with the areas of squares, rectangles, and triangles. Next comes the circle. If the diameter is d, the area is computed as

$$A = (1 - 1/7 - 1/14)d^2.$$

This means: the author uses Archimedes' estimate

$$\pi = 3 + 1/7,$$

just as Heron does.

The area of a circle segment is computed thus:

How is it with the bow? Let him add the arrow to the chord, both together, and multiply them into the half of the arrow and put them aside. And again let him take the half of the chord, multiply it into itself and divide it by 14, and the result let him add to the (added) standing (aside). The result is the area.

If s is the chord and a the arrow, the rule explained here is

(1) $$F = \tfrac{1}{2}(s + a)a + (1/14)s^2.$$

This is just the rule used by Heron for the case

$$2a \leqq s \leqq 3a.$$

The author adds:

But there are other methods.

Indeed, later on, in Chapter V, the author teaches another method, valid for segments less than a semi-circle. He first computes the diameter of the circle by means of the correct formula

$$d = (\tfrac{1}{2}s)^2/a + a,$$

next he calculates the area of the circle sector $ACBD$ (see Fig. 82) as

$$S = \tfrac{1}{2}d \cdot \tfrac{1}{2}b$$

where b is the arc. Subtracting the area of the triangle ABD

$$T = \tfrac{1}{2}h \cdot \tfrac{1}{2}s$$

he obtains the area of the segment

$$F = S - T.$$

Of course, the method is correct, but the author does not tell us how to compute the arc b.

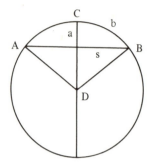

Fig. 82. Area of a circle segment

The volume of a rectangular block is computed as the product of length, breadth, and thickness.

The volume of a prism or cylinder is found by multiplying the height into the base area.

The volume of a pyramid or cone is found by multiplying the height into one-third of the base area.

A truncated pyramid is measured as the difference of two pyramids. Heron uses just the same method. The method is different from the Egyptian-Chinese rule

$$V=(1/3)h(a^2+ab+b^2)$$

but it gives just the same result.

Quadrangles are divided into five classes: squares, rectangles, rhombs, rhomboids, other quadrangles, just as in the Elements of Euclid and in the Heronic "Geometrika". The classification of triangles (right-angled, acute-angled, and obtuse-angled) is the same as in the Heronian writings.

In Chapter IV, the author teaches the calculation of the area of a triangle by means of the "Heronic" formula

$$T=\sqrt{s(s-a)(s-b)(s-c)}$$
$$s=\tfrac{1}{2}(a+b+c).$$

In Chapter V, the surfaces of spherical segments are calculated correctly, using the Archimedean estimate $\pi=3+1/7$.

Next follows a discussion of a passage in the Bible, in which it is said that a molten sea, round in compass, measures 10 cubits from brim to brim, while its circumference is said to be 30 cubits. The author tries to bring this into accord with the opinion of the "people of the world" who say that the circumference of a circle contains $3+1/7$ times the diameter. According to his interpretation, the difference is due to the thickness of the sea at the two brims. The author's opinion seems to be that the diameter of 10 cubits included the walls of the sea, while the circumference excluded them.

The Mishnat ha-Middot is important for the history of algebra, because it is closely related to the geometrical part of Al-Khwarizmi's Algebra. See on this subject a recent paper of E. Neuenschwander: Reflections on the Sources of Arabic Geometry, to appear in Journal for History of Arabic Science.

Chapter 7

Liu Hui and Aryabhata

The Chinese geometer Liu Hui lived in the third century A. D., and the Hindu astronomer Aryabhata lived in the beginning of the sixth century. I shall discuss their mathematical achievements in one chapter, for several reasons.

First, with Liu Hui we leave the domain of popular mathematics. His geometry is on the same level as the geometry of Euclid: he presents proofs of really difficult theorems. Aryabhata gives no proofs, but his trigonometry and mathematical astronomy are by no means elementary.

Secondly, Liu and Aryabhata both know the excellent estimate

$$\pi = 3.1416$$

which is probably due to Apollonios, and there are other signs of Greek influence in the work of Liu as well as in that of Aryabhata, as we shall see.

This chapter consists of two parts:
Part A: The Geometry of Liu Hui,
Part B: The Mathematics of Aryabhata.

Part A

The Geometry of Liu Hui

My information on Liu Hui comes from six sources:

1. Ho Peng-Yoke: Art. LIU HUI in Dictionary of Scientific Biography,
2. J. Needham: Science and Civilization in China, Vol. 3,
3. P. L. van Hee: Le classique de l'île maritime, Quellen und Studien Geschichte der Math. B 2, p. 255–280 (1932),
4. D. B. Wagner: An Early Chinese Derivation of the Volume of a Pyramid, Historia Mathematica 6, p. 164–188 (1979),

5. D. B. Wagner: Translation of Discussions on Circle-Mensuration by Liu Hui, etc. (unpublished),
6. D. B. Wagner: Liu Hui and Tsu Keng-chih on the Volume of a Sphere, Chinese Science 3, p. 59–79 (University of Pennsylvania, Philadelphia 1978).

According to van Hee and Ho Peng-Yoke, Liu Hui flourished in the middle part of the third century A. D. His commentary on the "Nine Chapters" has exerted a profound influence on Chinese mathematics for well over a thousand years. He wrote another, much shorter work: the "Sea Island Mathematical Manual", which has been translated into French by van Hee under the title "Le classique de l'île maritime", and which we shall discuss now.

The "Classic of the Island in the Sea"

The Chinese title of this treatise is *Hai-tao suan-ching*. It contains nine problems, all concerning the measurement of distances and heights. An English translation of the treatise is not available, so I shall translate some parts of the text from the French of van Hee. Since the text is perfectly clear and logical, there seems to be no danger in this procedure.

First Problem: The Island in the Sea

Someone wants to measure an island. Two poles are erected, both 30 feet high, one nearer to the island, the other farther away, in a distance of 1000 paces (1 pace = 6 feet). The pole that is farther away is exactly in one straight line with the first one and the island. If the eye looks towards the top of the first pole from the earth at a distance of 123 paces, it just sees the highest point of the island. If one places oneself in the same way 127 paces behind the other pole, one sees the peak of the island on the visual ray passing from the earth to the top of the second pole. To find the height of the island and its distance.

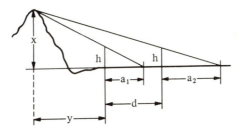

Fig. 83. The island in the sea

Liu Hui solves the problem by indicating exactly what multiplications, subtractions, divisions, and additions one has to perform in order to solve the problem. In Fig. 83 the unknown height is called x, its distance from the first pole y, the height of the poles h, their distance d, and the distances from the observers to the poles a_1 and a_2. The solution is correctly given as

$$x = \frac{hd}{a_2 - a_1} + h$$

$$y = \frac{a_1 d}{a_2 - a_1}.$$

Obviously, Liu Hui used the proportionality of sides in similar triangles, for it is easy to see that his x and y satisfy the two linear equations

$$\frac{x-h}{h} = \frac{y}{a_1}$$

$$\frac{x-h}{h} = \frac{y+d}{a_2}.$$

The problem is geometrical, but the solution is algebraic.

Second Problem: Height of a Tree

The second problem is similar to the first. One observes the top of a tree standing on a mountain by means of two poles erected in the plane. The eye of the observer is always on the earth. One of the observers also looks at the base of the tree. Liu Hui computes the height of the tree and the horizontal distance to the nearest observer. The solution is based on the same principles as before (see Fig. 84).

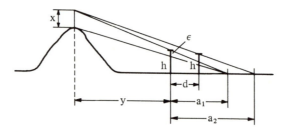

Fig. 84. The tree on the mountain

Third Problem: Square Town

An observer looks southward and sees a square town. He erects two stakes at a given distance, one to the west of the other, and he joins them by a chord at the height of his eyes. The eastern stake is just in one line with the N. E. and S. E. corners of the town. Our observer goes back 5 paces to the North and fixes the N. W. corner of the town, which he sees at a point of the chord in a given distance from the eastern end of the chord. He goes back north until he sees the N. W. corner just in one line with the western stake. Required: the side of the square and its distance to the stakes.

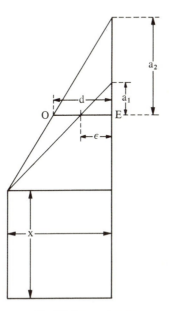

Fig. 85. Square town

If x is the required side and if the given distances d, ε, a_1, a_2 are as in Fig. 85, the solution is correctly given as

$$x = \frac{(a_1 - a_2)\varepsilon}{\varepsilon a_2 / d - a_1}.$$

I think this is sufficient to give the reader an idea of the contents of the treatise of Liu Hui. The remaining problems are similar, and the solution is always based on the same principles.

Still more interesting are certain parts of Liu Hui's commentary to the "Nine Chapters". In Wagner's opinion, some parts of the commentary usually ascribed to Liu Hui may be due to a later commentator. See Donald R. Wagner: Doubts Concerning the Attribution of Liu Hui's Commentary to the *Chiu-chang suan-shu*, Acta Orientalia 39, p. 199–212. Being unable to judge Wagner's arguments, I shall retain the traditional ascription and write "Liu Hui" whenever the author of the commentary is meant.

The Evaluation of π

According to the "Nine Chapters" and to Liu Hui, the area of any circle sector is equal to one quarter of the product of the arc and the diameter d. This holds also for the whole circle, so if we call the ratio of the circumference to the diameter π, the ratio of the area to $r^2 = (1/4)d^2$ will also be π.

In the "Nine Chapters", π was assumed to be 3. Liu Hui gives two improved estimates

$$\pi = 3.14 \quad \text{and} \quad \pi = 3.1416.$$

Both estimates are correct up to the last decimal.

How did Liu Hui obtain these estimates? Donald B. Wagner has kindly allowed me to make use of his translation of the relevant passages of Liu's commentary.

Problem 31 in Chapter 1 of the "Nine Chapters" reads:

A circular field has circumference 30 pu and diameter 10 pu. What is the area of the field?

In his commentary to this problem Liu Hui states that the area of the field is

$$71\frac{103}{157} \ (\text{square}) \, pu.$$

This implies that Liu started with the given circumference $C = 30$ and used the rule

(1) $$A = C^2/4\pi$$

with

$$\pi = \frac{157}{50} = 3.14.$$

Indeed, if this value of π is used, one obtains

$$A = \frac{900}{4} \times \frac{50}{157} = 225 \times \frac{50}{157} = 71\frac{103}{157}.$$

Another commentator, Li Shung-Feng[18] (7th century A.D.) makes the same calculation with $\pi = 22/7$, thus obtaining

$$A = 71\frac{13}{22}.$$

---

18 His name is sometimes pronounced as Li Ch'ung-Feng.

The value 22/7 was already known to the brilliant mathematician Tsu Ch'ung-Chih, who lived, according to Needham (Science and Civ. III, p. 101), from A. D. 430 to 501. He called the value 22/7 "inaccurate", and presented a more accurate value

$$\pi = \frac{355}{113} = 3.1415929\ldots.$$

This is excellent, for the modern value is

$$\pi = 3.1415926\ldots.$$

But let us return to Liu Hui. He first shows that $\pi = 3$ is too small, because 3 times the diameter of a circle is just the perimeter of an inscribed hexagon. He next shows that in a circle of radius r the area of an inscribed regular $2n$-gon (let's call this area A_{2n}) is just $\frac{1}{2} r$ times the perimeter of an inscribed regular n-gon:

(2) $A_{2n} = \frac{1}{2} r C_n$

As n becomes larger, "the loss is smaller", says Liu Hui. This means: A_{2n} approaches more and more the true area A of the circle.

Liu next gives an estimate of the difference $A - A_n$, and he shows that this difference tends to zero when n is increased infinitely. His proof is as follows:

Let AB be a side of the inscribed regular n-gon. Obviously, the area of the circle segment AEB is less than the area of the rectangle $ABCD$. Summing over the n sides of the polygon, one finds that the difference $A - A_n$ is less than the sum of the n rectangles. Since the height of the rectangles can be made arbitrarily small, the sum of their bases being bounded, the difference $A - A_n$ converges to zero. Of course, Liu Hui does not use the modern expression "convergence", but he expresses himself very clearly.

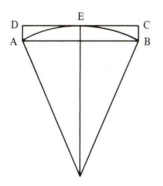

Fig. 86. Liu Hui's proof of convergence

From (2) Liu Hui obtains, by a correct passage to the limit, the relation between area and circumference of a circle of diameter d:

$$A = \tfrac{1}{2} d \cdot \tfrac{1}{2} C$$

Liu next explains his calculation of π. He takes a circle of diameter

$$2\,ch'ih = 20\,ts'un$$

and calculates by means of the Theorem of Pythagoras the sides of inscribed regular polygons of 6, 12, 24, 48, and 96 sides. Using (2), he obtains the areas

$$A_{96} = 313 + \frac{584}{625}\,ts'un$$

$$A_{192} = 314 + \frac{64}{625}\,ts'un.$$

He also forms the difference

(3) $$D = A_{192} - A_{96} = \frac{105}{625}\,ts'un.$$

It is clear that $2D$ is the area of the figure formed by the 96 rectangles $ABCD$. Adding this to the area A_{96}, Liu obtains

$$A_{96} + 2D = 314 + \frac{169}{625}\,ts'un$$

and he states correctly that his sum exceeds the area of the circle. The result of the calculation is equivalent to

(4) $$3.141024 < \pi < 3.142704.$$

Throwing away the fractions, Liu concludes that 314 is the "determined portion" of the area of the circle. Dividing the area by the radius and doubling the result, he obtains the "number of the circumference"

$$6\,ch'ih\ 2\,ts'un\ 8\,fen\ =\ 6.28\,ch'ih$$

$$(1\,ch'ih\ =\ 10\,ts'un$$
$$1\,ts'un\ =\ 10\,fen).$$

After a digression concerning the inscribed and circumscribed square of the circle, Liu states that his estimate 314 is a bit too small, and he proceeds to improve on it. He repeats that the difference D is $105/625\,ts'un$, and he says that it is necessary to take 36 of these 105 parts and add them to the area of the 192-gon to get the area of the circle:

$$A = 314 \frac{64}{625} + \frac{36}{625} = 314 \frac{4}{25} \; ts'un.$$

If one would use Heron's method, one would have to take

$$\frac{1}{3} \cdot \frac{105}{625} = \frac{35}{625}$$

instead of 36/625. But Heron says and even proves that this approximation is a little too small. Therefore it is not unreasonable to replace 35 by 36, thus obtaining Liu's admirable estimate 3.1416.

From this result Liu deduces:

$$\frac{\text{Area of Circle}}{\text{Area of Circumscribed Square}} = \frac{3927}{5000}$$

and

$$\frac{\text{Diameter}}{\text{Circumference}} = \frac{1250}{3927}.$$

Finally, Liu states that the above method is still approximate. One should calculate the side of the 1536-gon, obtain the area of the 3072-gon, and neglect the minute fraction, says he.

Liu does not say what the result of this calculation would be. If one calculates the area of the 3072-gon by modern methods and rounds off to 5 decimals, one obtains the estimate

$$\pi = 3.14159$$

which Ho Peng-Yoke ascribes to Liu Hui.

By the way: if one expands Liu's estimate 3.1416 into a continuous fraction, one obtains

$$\frac{31416}{10000} = 3 + 1/[7 + 1/(16 + \ldots)].$$

Neglecting the +..., one obtains Tsu Ch'ung Chih's estimate

$$3 + 1/(7 + 1/16) = \frac{355}{113}$$

which happens to be even more accurate than Liu's.

The Volume of a Pyramid

In Chapter 2 we have already discussed Liu's proof of the rule for computing the volume of a truncated pyramid on a square base

(5) $V=(1/3)h(a^2+ab+b^2)$.

In this proof the volume of a pyramid on a rectangular base

(6) $V=(1/3)hab$

is presupposed. To be more precise, the rule (6) is presupposed only for the special case of a *yang-ma*, that is, a pyramid on a rectangular base in which one of the edges is perpendicular to the base plane (Fig. 87).

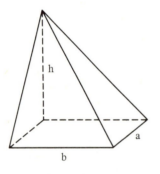

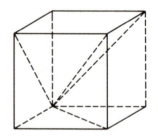

Fig. 87. *yang-ma*   Fig. 88. Division of a cube into three pyramids

In his commentary on Chapter 5 of the "Nine Chapters", Liu Hui gives a proof of rule (6). The proof is not a strict proof in the sense of Greek geometry, but it is completely convincing. An English translation of the proof has been given by Wagner in his paper in Historia Mathematica 6. Here I shall give an outline of the main idea.

Liu first considers the case $h=a=b$. He shows that a cube can be divided into three congruent *yang-ma*'s of this kind (see Fig. 88). It follows that each of the *yang-ma*'s is equal to one third of the cube. So in this case (6) is obvious.

Liu now passes to the general case. If a *yang-ma* has dimensions a, b, h as shown in Fig. 87, it can be fitted together with a *pieh-nao* (a tetrahedron) so as to form a prisma *(ch'ien-tu)*, as shown in Fig. 89. Now let us put

C = Volume of the *ch'ien-tu* $ABDCEF$,
Y = Volume of the *yang-ma* $BDFEC$,
P = Volume of the *pieh-nao* $BACE$.

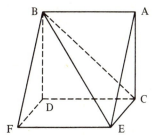

Fig. 89. Tetrahedron and pyramid fitted together

The volume of the prisma is well known

$$C = \tfrac{1}{2}abh$$

Therefore, to prove that $Y = (1/3)abh$, it is only necessary to show that $Y = 2P$. To prove this, Liu Hui divides up the *pieh-nao* and the *yang-ma* as in Fig. 90 and Fig. 91 respectively. The plane $IJLM$ in Fig. 90 is perpendicular to AC and halves AC, and the plane $HIKL$ in Fig. 91 is perpendicular to BD and halves BD. The other dividing planes are vertical.

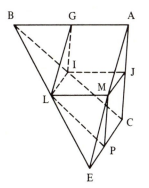

Fig. 90. Division of the *pieh-nao*   Fig. 91. Division of the *yang-ma*

We have now the following situation:

The *pieh-nao* of Fig. 90 is divided into two *ch'ien-tu* (wedges) and two *pieh-nao*.

The *yang-ma* of Fig. 91 is divided into one rectangular block and two wedges and two *yang-ma*.

Now the sum of the two wedges in Fig. 90 is just one half of the sum of the volumes of the block and the two wedges in Fig. 91. These pieces, which are in the ratio of 1:2, are called by Liu Hui the *determined volumes*. Taken together, they occupy just three quarters of the prisma shown in

Fig. 89. The remaining parts, which are still undetermined, are two *pieh-nao* pieces similar to the original *pieh-nao,* and two *yang-ma* pieces similar to the original *yang-ma.*

Now the process is repeated for the smaller *pieh-nao* and *yang-ma* pieces. Again, we have a certain number of determined pieces which are in the ratio of 1:2, and which together form 3/4 of the volumes remaining in the first step. What remains is only 1/4 of 1/4 of the prisma shown in Fig. 89.

For us, it is clear that the process converges. The remaining undetermined pieces are together less than an arbitrarily given volume v, however small. Liu Hui expresses this convergence as follows:

The smaller they are halved, the finer are the remaining (parts). The extreme of fineness is called "subtle". That which is subtle is without form. When it is explained in this way, why concern oneself with the remainder?

I think, this is a completely convincing proof of the desired result

$$V = (1/3)hab.$$

I don't quite see why Wagner says that Liu Hui has difficulty expressing the idea of carrying the process to the limit. In my opinion, when Liu says that the halving is continued until the remainder is extremely fine and subtle so that it can be disregarded, this is just as clear as when Euclid says that the remainder can be made less than any given volume, or when we say that it can be made less than any given ε.

Liu Hui and Euclid

In Book 13 of the Elements, Euclid proves:

XII 5. Pyramids on triangular bases having the same height are to each other as the bases.

If this is proved, it is easy to show that every pyramid on a triangular base is equal to one third of a prism having the same height and the same base, for the prism can be divided into three pyramids which have, two by two, equal bases and equal heights and are therefore equal. From a testimony of Archimedes ("On the Sphere and Cylinder", Introduction) we know that the proof of Euclid's final result is due to Eudoxos.

Now how does Euclid prove XII 5? Euclid's pyramid on a triangular base is just a *yang-ma,* or tetrahedron. Euclid divides it, just as Liu Hui does, into two equal prisms

$$BZK\text{-}EH\Theta \quad \text{and} \quad HZ\Gamma\text{-}\Theta K\Lambda$$

and two congruent pyramids

$$AEH\Theta \quad \text{and} \quad \Theta K\Lambda\Delta$$

(see Fig. 92). The two prisms are together more than half of the original pyramid. Now the two similar pyramids are again divided into pieces in the
same way, and so on.

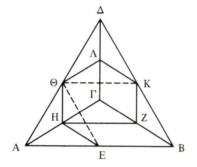

Fig. 92. Division of a tetrahedron according to Euclid

Now comes the crux. Euclid makes use of proposition X 1, which
reads:

Let A and Γ be comparable magnitudes, and A larger than Γ. If a piece larger than one
half of A is taken away from A, and from the remainder a piece larger than one half of it, and
so on, then at some time the remainder will be a magnitude less than Γ.

According to Archimedes, a lemma equivalent to this proposition had
already been used by Eudoxos in his determination of the volume of the
pyramid. See O. Becker: Eudoxos-Studien I, Quellen und Studien Geschichte der Math. B 2, p. 311–333, and also E. Neuenschwander: Die ste-
reometrischen Bücher der Elemente Euklids, Archive Hist. of Exact Sc. 14,
p. 109–118.

In his proof of XII 5, Euclid takes away from the pyramid two prisms
which together are more than half the pyramid. From the remainder he
again takes away more than half, and so on, until the remainder is less than
any given volume v. Just so, Liu Hui takes away three quarters from the
prism shown in Fig. 89, and from the remainder he again takes away three
quarters, and so on, until the remainder is completely neglible. Liu's proof
is very much similar to that of Euclid, and it is based on just the same division of a tetrahedron into two prisms and two smaller tetrahedra.

This ressemblance alone is not sufficient to conclude that Liu Hui was
influenced by Greek sources, but there are still other indications. Archimedes obtained his lower estimate $3 + 10/71$ of π from the circumference
of a polygon of 96 sides or, what amounts to the same, from the area of a
polygon of 192 sides. Liu uses the same area.

It is very remarkable that in the case of the pyramid as well as in the
case of the circle Liu's methods are essentially the same as the methods of
Eudoxos and Archimedes.

Another indication of Greek influence is Liu's estimate of π, which is
the same as Aryabhata's. The origin of this estimate will be discussed in
Part B of the present chapter.

Liu Hui on the Volume of a Sphere

In Chapter 2 we have seen that in the Chinese "Nine Chapters" the volume of a sphere with diameter d was calculated according to the rule

(7)
$$S = \frac{9}{16} d^3.$$

In his commentary, Liu Hui says:

The creator of the method used the proportions: circumference 3, diameter 1. If it is supposed that the area of a circle occupies 3/4 of the area of a (circumscribed) square, then a cylinder also occupies 3/4 of a (circumscribed) cube.

So far everything is clear and simple. Next Liu conjectures that the creator of the method supposed that the sphere also occupied 3/4 of the circumscribed cylinder. From this supposition it would follow that the sphere occupies 9/16 of the cube.

But, says Liu, this way of thinking is incorrect. He considers two cylinders, both inscribed in one and the same cube, with perpendicular axes (see Fig. 93). Their intersection is a solid which Liu Hui calls "lid of a square box". To visualize this solid, it is convenient to divide the cube into eight smaller cubes, and to depict the situation in one of these (see Fig. 94).

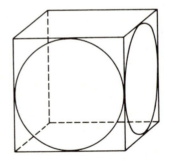

Fig. 93. The box-lid, an intersection of two cylinders

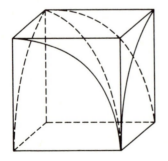

Fig. 94. An eigth part of the box-lid

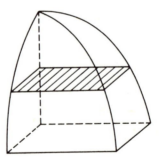

Fig. 95. Plane section of an octant of the box-lid

The part of the box-lid contained in one of the smaller cubes is shown in Fig. 95.

A. Youschkevitch has drawn my attention to the fact that the "box-lid" was already known to Archimedes, who proved, in his treatise "The Method", that the "box-lid" equals 2/3 of the cube (see my "Science Awakening I", p. 216). Youschkevitch feels that this fact furnishes a strong argument in favour of the hypothesis of a Greek influence on Liu Hui.

Liu Hui (or whoever wrote this part of the commentary) shows, by a very ingenious proof, that the inscribed sphere has to the box-lid the same ratio as a circle to its circumscribed square. If the sphere is called S and the box-lid B, we may write this result as

(8)
$$S = \frac{\pi}{4} B.$$

Now B is less than one of the cylinders C, so we have

$$S < \frac{\pi}{4} C.$$

Hence, if we put S equal to $(\pi/4)C$, the result will be too large. On the other hand, if we replace π by 3, the result will be too small. Liu Hui concludes that the two errors partly compensate each other. "Therefore", he says, "the proportions 9 and 16 are accidentally close to the truth, but the sphere is still too large", i.e. the volume of the sphere calculated by (7) is too large.

In order to prove that S is to B as a circle is to its circumscribed square, Liu uses a very remarkable principle, closely related to what we call the "Principle of Cavalieri". This latter principle, applied to volumes, says that two solids of equal height have the same volume, if plane sections at equal heights always have the same areas. See Bonaventura Cavalieri: Geometria indivisibilibus continuorum nova quadam ratione promota (Bologna 1635).

Cavalieri's principle can be proved by means of Eudoxos' "Exhaustion Method". More precisely, one can show that the principle holds for all measurable solids in the sense of Jordan. For Jordan's definition of measurability see Camille Jordan: Cours d'Analyse I, 3rd ed. (1909), p. 28.

Cavalieri was not the first to formulate this principle. The Chinese mathematician Tsu Keng-Chih, who lived in the late 5th century A. D., expressed the same idea in the form of a poem. In Wagner's translation the poem reads:

> If volumes are constructed of piled-up blocks,
> And corresponding areas are equal,
> Then the volumes cannot be unequal.

The commentator Li Huang (died 1812) emended the sign *ch'i*, "block", to the similar sign *mi*, "area". As Wagner remarks, this emendation fits very well with Tsu Keng-Chih's actual method.

Liu Hui, in his calculations of volumes, uses a similar principle. Wagner enunciates his principle thus:

If an object with circular cross-section is inscribed in an object with square cross-section, and every circular cross-section is inscribed in the corresponding square cross-section, then the ratio of the volumes of the objects is equal to the ratio of the areas of a circle and a circumscribed square.

Liu himself does not (as far as I know) explicitly formulate this principle, but he clearly uses it in his proofs. For instance, the sphere is inscribed in the box-lid, and the plane sections of the box-lid are squares (see Fig. 95), and the plane sections of the sphere are the inscribed circles of these squares. From this, Liu concludes that the sphere is to the box-lid as a circle is to its circumscribed square, i.e. as π is to 4.

Liu was not able to calculate the volumes of the box-lid and of the sphere. He concludes with a witty kind of poem, translated by Wagner thus:

> Look inside the cube
> And outside the box-lid;
> Though the diminution increases,
> It doesn't quite fit.
> The marriage preparations are complete;
> But square and circle wrangle,
> Thick and thin make treacherous plots,
> They are incompatible.
> I wish to give my humble reflections,
> But fear what I will miss the correct principle;
> I dare to let the doubtful points stand,
> Waiting
> For one who can expound them.

Donald Wagner, in his paper 6., cited at the beginning of the present chapter, considers the possibility that the proof of (8) and the poem are not due to Liu Hui but to a later commentator, perhaps to Tsu Ch'ung-Chih, who wrote a commentary to the "Nine Chapters", now lost.

The volume of the box-lid was determined by Tsu Ch'ung-Chih's son Tsu Keng-Chih. He proved by means of Cavalieri's principle that the box-lid is 2/3 of the circumscribed cube:

$$(9) \qquad\qquad B = (2/3)\, d^3 .$$

His proof of this result is reproduced in Donald Wagner's paper 6. in "Chinese Science", Vol. 3.

Combining (8) and (9), the younger Tsu finds the volume of the sphere

$$(10) \qquad\qquad S = (\pi/6)\, d^3 .$$

Tsu Keng-Chih now replaces π by 3 and says:

Therefore I have said that a sphere occupies one-half of a cube.

He expresses his satisfaction by another poem:

> The proportions are extremely precise,
> And my heart shines.

Chang Heng copied the ancient,
Smiling on posterity;
Liu Hui followed the ancient,
Having no time to revise it.
Now what is so difficult about it?
One need only think.

According to Wagner, the quotation from Tsu Keng-Chih probably ends here. The whole quotation, including the proof and the two poems, has been translated by Wagner from Li Ch'un-Feng's commentary on the "Nine Chapters".

Part B

The Mathematics of Aryabhata

The Aryabhatiya of Aryabhata[19] is an astronomical treatise of great perfection. It is based on a pre-Ptolemy Greek theory of epicycles and eccenters, and on Greek trigonometry. As R. Billard has shown in his book "L'astronomie indienne", the constants of Aryabhata's astronomical system were determined by means of very accurate observations made in India about A. D. 510.

The second chapter of the Aryabhatiya is entitled "Ganita", which means Calculation or Mathematics. It is a popular text, comparable with the Chinese "Nine Chapters" or the work ascribed to Heron. It contains prescriptions for solving mathematical problems, without proofs.

We shall now discuss some of Aryabhata's rules of calculation. For a thorough discussion of the whole chapter "Ganitapada" see K. Elfering: Die Mathematik des Āryabhata I (Verlag Fink, München 1975).

Area and Circumference of a Circle

The first half-verse of Aryabhatiya II 7 reads in the translation of Clark:

Half of the circumference multiplied by half to the diameter is the area of a circle.

This is the well-known rule

(1) $$A = \tfrac{1}{2} C \cdot \tfrac{1}{2} d$$

which is also found in Greek and Chinese sources.

The second half-verse deals with the sphere. Since its interpretation is doubtful, I shall not discuss it here.

---

19 English translation by W. E. Clark: The Āryabhatiya of Āryabhata, Chicago 1930. Text and English translation by K. S. Shukla and K. V. Sarma, Indian National Science Academy, New Delhi 1976.

II 10 reads:

Add 4 to 100, multiply by 8, and add 62000. The result is approximately the circumference of a circle of which the diameter is 20000.

This means that Aryabhata's estimate of π is

$$\pi = \frac{62832}{20000} = 3.1416.$$

Bhaskara II, also called Bhaskaracarya, writes the same fraction in the reduced form $3927/1250$ (see H. T. Colebrooke: Algebra with Arithmetic and Mensuration from the Sanskrit of Brahmegupta and Bháscara, London 1817, p. 87).

We have seen that Liu Hui has the same value of π, and that he writes it in the same reduced form. Liu adds that his result is only approximate, and Aryabhata too says that his result is approximate. I guess that the statements of Liu Hui (3rd century), Aryabhata (6th century), and Bhaskara II (12th century) all are derived from one and the same source.

In what follows, I shall try to identify this source.

Aryabhata's Table of Sines

The verses II 11–12 of the Ganitapada deal with the calculation of Sines. The Hindu Sine is, in modern notation, defined by

$$\mathrm{Sin}\,\varphi = R\sin\varphi$$

where R is an arbitrary radius. Following Aryabhata, we shall express all angles and also their Sines in minutes of arc. In Aryabhata's table of Sines (Aryabhatiya I 10) R is chosen as $3438'$. The advantage of this choice is that for small angles the Sine is nearly equal to the angle.

To find a value of R satisfying this condition, one has to know an estimate of π beforehand. For small angles φ (expressed in minutes) one has

$$R\sin\varphi \sim R \cdot \frac{\pi}{180 \times 60}\,\varphi$$

and if one wants to make this equal to φ, one has to make R equal to

(2) $$R = \frac{10800'}{\pi}$$

Substituting Aryabhata's value of π, one obtains

$$R = \frac{10800'}{3.1416} = 3437'.7\ldots$$

which may be rounded up to Aryabhata's value $3438'$.

In verse II 11 Aryabhata indicates, in a very concise form, how the Sines may be computed, starting with a square and an equilateral triangle inscribed in a circle. The translator Clark was not able to reconstruct Aryabhata's method, but the commentator Bhaskara I (A. D. 629) has given a clear explanation (see K. S. Shukla, Āryabhatīya of Āryabhata, p. 45–51).

Bhaskara divides an arc of 90° in a circle of radius 3438' first into 6, next into 12, and finally into 24 parts. He computes the length of 24 chords, and by halving the chords he obtains the 24 Sines of Aryabhata. The fact that Bhaskara uses chords seems to indicate that the Hindu tables of Sines were originally derived from Greek tables of chords.

In verse II 12 Aryabhata indicates another method of computing Sines. Let us denote his twenty-four Sines by $R_1,\ldots,R_{24}$. We have, by definition,

$$R_n = \operatorname{Sin} n\alpha = R \sin n\alpha$$

with

$$a = 90°/24 = 225'.$$

Aryabhata's second method starts with $R_1 = 225'$. From this first Sine, the others are found by means of the rule

(3) $$R_{n+1} - R_n = R_1 - C(R_1 + \ldots + R_n)$$

with a constant C. According to modern trigonometry, this rule is correct if one makes C equal to

$$C = 2(1 - \cos\alpha).$$

For $\alpha = 225'$ one would obtain

$$C = 0.00428.$$

Aryabhata prescribes to divide the sum $R_1 + \ldots + R_n$ by $R_1 = 225'$, which means that he makes C equal to

$$\frac{1}{225} = 0.00444$$

which is not correct. He should have prescribed a division by 233.6 instead of a division by 225. In the beginning, where $R_1 + \ldots + R_n$ is still small, it makes little difference whether one divides by 225 or 233.6, but at the end the differences $R_{n+1} - R_n$, calculated by (3), are systematically too small.

The correct rule (3) can be justified by means of Proposition 22 of Archimedes' treatise "On the Sphere and Cylinder I". Namely: if one multiplies both sides of (3) by 2, one obtains on the right a sum of chords in a circle, and the proposition 22 tells us how to express this sum of chords by means of a simple proportion. In the translation of Th. Heath (The Works of Archimedes, p. 29) this proposition reads:

If a polygon be inscribed in a segment of a circle LAL' so that all its sides excluding the base are equal and their number even, as $LK...A...K'L'$, A being the middle point of the segment, and if the lines BB', CC', ... parallel to the base LL' and joining pairs of angular points be drawn, then

$$(BB'+CC'+...+LM):AM=A'B:BA,$$

where M is the middle point of LL' and AA' is the diameter through M.

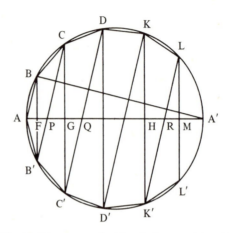

Fig. 96. Diagram to Archimedes' proposition 22

The drawing (Fig. 96) is made for the case of a circle segment LAL' larger than a semi-circle, but the theorem is valid for all segments. If one applies it to a segment not exceeding a semi-circle, one can easily derive (3).

In Shukla's commentary to II 12, on p. 53–54 of his edition of the Aryabhatiya, one finds two other proofs of the rule (3), with the correct factor $2(1-\cos\alpha)$. The first proof, due to Shukla himself, makes use of modern trigonometry. The second is a geometrical proof in classical Greek style, due to Nilakantha, who wrote his commentary to the Aryabhatiya about A.D. 1500 (see p. xlv of Shukla's edition).

For $n=1$ the formula (3) yields

$$R_2=R_1+R_1-CR_1$$

or

(4) $$R_2-R_1=R_1-CR_1.$$

In Aryabhata's table of Sines the first Sine R_1 is 225, and the difference R_2-R_1 is 224. If one substitutes these values into (4) and calculates C, one finds $C=1/R_1$, so (4) becomes

(5) $$R_2-R_1=R_1-R_1/R_1.$$

This equation (which is correct only if one neglects fractions of minutes in R_1 and R_2) is formulated by Aryabhata in the first sentence of II 12:

The first Sine divided by itself and then diminished by the quotient gives the second Sine difference (translation of Shukla and Sarma).

In the second sentence of II 12, Aryabhata applies the same inaccurate value $C = 1/R_1$ to formulate the rule (3) thus:

The same first Sine diminished by the (sum of) the quotients obtained by dividing the preceding Sines by the first Sine gives the remaining Sine differences.

From the preceding analysis of Aryabhata's text and Nilakantha's commentary we may draw two conclusions:

1. A geometrical justification of the rule (3) was known to the Hindu astronomers.

2. This justification was not invented by Aryabhata, for otherwise he would not have used his inaccurate value $C = 1/R_1$.

On the Origin of Aryabhata's Trigonometry

Aryabhata's three assertions II 10–12 belong together. In II 10 the value of π is presented, which is a necessary prerequisite for the calculation of the Sines indicated in II 11. In II 12 a linear equation between the Sines is presented, which can be used as a check on the calculation.

To this set of three closely related assertions we must associate the Sine table itself (Aryabhatiya I 10), which may be computed by means of II 11 and checked by means of II 12. Thus, the stanzas I 10 and II 10–12 form together one logically connected theory.

Can we say more about the origin of this theory? Yes, we can. In a paper entitled "The Chord Table of Hipparchos and the Early History of Greek Trigonometry", Centaurus 18, p. 6–28 (1974), C. G. Toomer has shown that the chord table of Hipparchos was a table of chords in a circle of radius $R = 3438'$. This is just Aryabhata's value. In my opinion, Toomer is justified in concluding that Aryabhata's table of Sines was derived from Hipparchos' table of chords by halving the chords.

We may advance one more step beyond Toomer's conclusion. In Aryabhata's text, we have not only a table of Sines, but a whole logically connected theory. So we must ascribe this theory either to Hipparchos or to a mathematician before Hipparchos.

I have serious doubts whether Hipparchos himself was able to invent a completely new theory. He was famous for his excellent observations, but no mathematical discovery or paper has been ascribed to him, as far as we know. He is said to have encouraged the mathematicians to investigate why the theories of epicycles and of eccenters both lead to the same phenomena (Theon of Smyrna, ed. Hiller, p. 166). For a good mathematician, the proof would be very easy. According to Ptolemy (Almagest IV 11) Hipparchos made several errors in his lunar theory. So it seems much more probable that the trigonometrical theory used by Hipparchos was invented by an earlier mathematician.

This mathematician must have had an estimate of π, more accurate than Archimedes' value 22/7. For if one calculates R from our formula (2) with $\pi = 22/7$, one obtains

$$R = 10\,800 \times \frac{7}{22} = 3\,436.4$$

and not 3 438.

So the originator of the theory connected with the Sine table must have been an able mathematician living between Archimedes and Hipparchos. Who was it?

There is just one excellent mathematician who lived in this time, namely Apollonios of Perge (ca. 200 B.C.). From the testimony of Eutokios (Commentary on Archimedes, Archimedis Opera III, ed. Heiberg, p. 258) we know that Apollonios wrote a paper entitled "Rapid Delivery" ('Ωκυτόκιον), in which improved limits for π were established. The testimony of Eutokios reads:

But one must know that Apollonios of Perge in his treatise Okytokion has demonstrated the same thing (as Archimedes) through other numbers leading to a more accurate approximation. This appears to be more exact, but it is not useful for the purpose of Archimedes, for we have said that his purpose in this book was to find an approximation useful for daily life.

The words "we have said" refer to an earlier statement of Eutokios (ed. Heiberg, p. 228):

As Herakleides says in his "Life of Archimedes" this book (the "Measurement of the Circle", by Archimedes) is necessary for the needs of daily life.

So Apollonios satisfies all conditions we have specified for the inventor of the theory underlying Aryabhata's verses I 10 and II 10–12. Moreover, he is the only know mathematician satisfying these conditions.

If this is accepted, we are bound to conclude that the estimate $\pi = 3.1416$ is due to Apollonios. This conclusion is confirmed by a statement in the Lilavati of Bhaskaracarya. It reads in the translation of H.T. Colebrooke (Algebra with Arithmetic and Mensuration from the Sanskrit of Brahmegupta and Bháscara, London 1817, p. 87):

When the diameter of a circle is multiplied by three thousand nine hundred and twenty-seven and divided by twelve hundred and fifty, the quotient is the nearly precise circumference, or multiplied by twenty-two and divided by seven, it is the gross circumference adapted to practice.

The last words "adapted to practice" correspond exactly to the words of Eutokios "useful for daily life". In the text of Eutokios as well as in that of Bhaskara the value 22/7 is said to be good enough for practical needs. In both texts a more accurate approximation is mentioned: in the text of Eutokios it is the approximation of Apollonios, and in the text of Bhaskara it is the estimate

$$\pi = \frac{3\,927}{1\,250} = 3.1416.$$

The correspondence between the two texts is so close that we are bound to conclude that they go back to a common source, and hence that the estimate of π is due to Apollonios.

Note that Liu Hui has exactly the same estimate, and that he writes it in the same reduced form as

$$3927/1250.$$

I suppose that Liu Hui, Aryabhata, and Bhaskara II all had this estimate of π from Apollonios.

Apollonios and Aryabhata as Astronomers

Apollonios was not only a great geometer, but also a great astronomer, and Aryabhata was mainly an astronomer. Now astronomy and mathematics are closely connected: trigonometry is an indispensable tool of Epicycle Astronomy. So if we want to see the work of Apollonios and that of Aryabhata in a fitting historical perspective, we cannot restrict ourselves to pure mathematics.

Aryabhata's astronomical system is based on the theory of epicycles and eccentric circles. Now the one who developed this theory in its mature from was Apollonios of Perge. So it seems evident that Aryabhata was influenced by Apollonios, not only in mathematics but also in the domain of astronomy.

Aryabhata himself acknowledges his indebtedness to his predecessors. His own words are:

IV 49. By the grace of Brahma the precious sunken jewel of true knowledge has been brought up by me from the ocean of true and false knowledge by means of the boat of my own intellect.

This beautiful image seems to indicate that Aryabhata used his own intellect not to discover the whole theory anew, but to distinguish between true and false knowledge, and to expose the theory in a consistent form. His testimony is humble and proud at the same time. The precious jewel existed, but it was sunken, and he rescued it. We now know that he used not only his intellect, but also accurate observations of the sun, the moon, and the planets, made about A. D. 510.

According to Ptolemy (Almagest XII,1) Apollonios proved two theorems on stationary points and retrograde motion of the planets, one being based on the epicycle hypothesis and the other on the hypothesis of eccentric circles. He was fully aware of the equivalence of the two hypotheses. As Neugebauer rightly remarks "the brilliant mathematical treatment of both cinematic models as one common structure is undoubtedly the work of Apollonius"[20].

Apollonios is said to have made a special study of the motion of the moon and to have computed lunar tables. For a careful discussion of the ancient testimonies see again Neugebauer[20], p. 262–263. Now for comput-

---

20 O. Neugebauer: A History of Ancient Mathematical Astronomy I, p. 263. See also Neugebauer's paper "The Equivalence of Eccentric and Epicyclic Motion According to Apollonius, Scripta Mathematica 24, p. 5–21 (1959).

ing lunar longitudes by means of epicycles or eccenters one needs trigonometry. Thus, our earlier conclusion that Apollonios composed a table of chords is confirmed once more. I suppose that he used chords in a circle of radius $R = 3438'$ for his successor Hipparchos had a table of chords based on the same radius, as we have seen earlier.

Why did Apollonios (or whoever invented this system) use such a complicated radius? A table with $R = 60$, as used by Ptolemy, is much easier to compute. We can find a possible answer by looking at Aryabhata's method for computing positions of the sun and the moon. In his computations the Sines of small angles are replaced by the angles themselves. This is only possible if one chooses $R = 3438'$.

So the pieces of our jig-saw puzzle fit together nicely, and we may safely conclude that the essential features of Aryabhata's astronomical system, including his trigonometry, are due to Apollonios of Perge.

Let us now return to Aryabhata's mathematics.

On Gnomons and Shadows

A gnomon is a vertical rod casting its shadow on a horizontal plate. Aryabhatiya II 14 reads:

Add the square of the height of the gnomon to the square of its shadow. The square root of this sum is the radius of the "sky-circle".

The reading of the word *khavṛtta,* which means sky-circle, is uncertain, but it is quite clear that the hypotenuse of the right-angled triangle formed by the gnomon and its shadow is meant. The calculation of this hypotenuse is a straightforward application of the Theorem of Pythagoras.

According to the commentator Bhaskara I, the purpose of the calculation of the hypotenuse is, to find the Sines of the angles of the triangle. From the Sines one can determine the angles by means of a table of Sines.

For the next stanza I shall use the translation and commentary of Shukla:

II 15. Multiply the distance between the gnomon and the lamp-post (the latter being regarded as base) by the height of the gnomon and divide (the product) by the difference between (the heights of) the lamp-post (base) and the gnomon. The quotient (thus obtained) should be known as the length of the shadow measured from the foot of the gnomon.

In Fig. 97 let AB be the lamp-post, CD the gnomon, and E the point where AC and BD produced meet. Then DE is the shadow cast by the gnomon due to light from the lamp at A.

Let FC be parallel to BD. Then comparing the similar triangles CDE and AFC, we have

$$DE = \frac{FC \times CD}{AF} = \frac{BD \times CD}{AB - CD}.$$

Hence the rule.

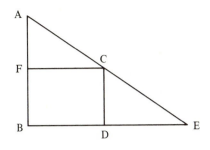

Fig. 97. Shadow of a gnomon

In the text, the height of the lamp-post is called *bhujā,* and the horizontal side *B E* of the triangle *A B E koti.* Quite generally, the terms *bhujā* and *koti* denote two perpendicular sides of a right-angled triangle, just as the terms *kou* and *ku* in the Chinese "Nine Chapters".

The next stanza reads in Clark's translation:

II 16. The distance between the ends of the two shadows multiplied by the length of the shadow and divided by the difference in length of the two shadows gives the *koti.* The *koti* multiplied by the length of the gnomon and divided by the length of the shadow gives the length of the *bhujā.*

Clark explains the calculation thus:

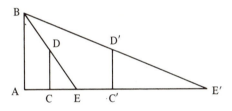

Fig. 98. Shadows of a gnomon in two positions

A B is the *bhujā*
A E is the *koti*
C D is the gnomon in its first position
C′ D′ is the gnomon in its second position
C E and *C′ E′* are the first and second shadows,

$$AE = \frac{CE \times EE'}{C'E' - CE'}$$

$$AB = \frac{AE \times CD}{CE}.$$

The method of calculation is similar to the methods employed by Liu Hui in his "Classic of the Island in the Sea". As we have seen, Liu and Aryabhata also have the same value of π. A common origin seems to be the most likely explanation of these similarities.

These excerpts may suffice to give the reader an idea of the geometry of Aryabhata. We now come to his algebra.

Square Roots and Cube Roots

In II 4 Aryabhata teaches the extraction of square roots, in II 5 of cube roots. He instructs the reader to determine one decimal after the other. His method is based on the identities

$$(a+b)^2 = a^2 + 2ab + b^2$$
$$(a+b)^3 = a^3 + 3a^2b + 3ab^2 + b^3.$$

Aryabhata's procedure is equivalent to the method in the "Nine Chapters", but his order of operation is slightly different. In the case of the square root, he starts with a one-digit approximation a (a multiple of 1, or 10, or 100, as the case may be), he subtracts a^2 from the given number N, he estimates b by dividing the difference $N - a^2$ by $2a$, he subtracts $2ab$ and b^2, etc. In the case of the cube root the method is similar. For more details see the book of Elfering: Die Mathematik des Aryabhata I (München 1975), p. 51–65.

Arithmetical Progressions and Quadratic Equations

Stanzas 18 and 20 in Chapter II of the Aryabhatiya deal with arithmetical progressions. Stanza 19 reads in the translation of Clark:

II 19. The desired number of terms minus one, halved, plus the number of terms which precedes, multiplied by the common difference between the terms, plus the first term, is the middle term. This multiplied by the number of terms desired is the sum of the desired number of terms.

Let a be the first and f the last term of an arithmetical progression, and d the common difference. Aryabhata gives a rule to calculate the sum of n successive terms, extending from the $(p+1)$th to the $(p+n)$th term. Tacitly supposing n to be odd, he calculates the middle one of the n terms as

$$M = a + \left(\frac{n-1}{2} + p\right)d.$$

Multiplying this middle term by n, he finds the sum of the n terms. Of course, this result is also valid if the number of terms is even.

In the last sentence of the text, a rule is given to find the sum of the whole progression. If N is the total number of terms, the rule says

$$S = (a+f)N/2.$$

In the next stanza, the sum S, the difference d, and the first term a of an arithmetical progression are given, and it is required to find the number of terms N. This problem leads to quadratic equation for N:

$$N\left(a + \frac{N-1}{2}d\right) = S.$$

The solution is correctly given as

$$N = \frac{1}{2}\left(\frac{\sqrt{8dS+(d-2a)^2}-2a}{d} + 1\right)$$

or in Aryabhata's own words

II 20. Multiply the sum of the progression by eight times the common difference, add the square of the difference between twice the first term and the common difference, take the square root of this, subtract twice the first term, divide by the common difference, add one, divide by two. The result will be the number of terms.

As Elfering notes, an equivalent rule of computation has been presented by Diophantos in his treatise "On Polygonal Numbers". See Th. Heath: Diophantos of Alexandria (second edition, Cambridge 1910), and K. Elfering: Die Mathematik des Āryabhata I, p. 130. Once more we see that Aryabhata was probably influenced by Greek mathematics.

Index

B. L. van der Waerden

Algebra 1

8. Auflage der Modernen Algebra. 1971.
XI, 272 Seiten
(Heidelberger Taschenbücher, Band 12)
ISBN 3-540-03561-3

Inhaltsverzeichnis: Zahlen und Mengen. – Gruppen. – Ringe und Körper. – Vektorräume und Tensorräume. – Ganzrationale Funktionen. – Körpertheorie. – Fortsetzung der Gruppentheorie. – Die Theorie von Galois. – Ordnung und Wohlordnung von Mengen. – Unendliche Körpererweiterungen. – Reelle Körper.

B. L. van der Waerden

Algebra II

Unter Benutzung von Vorlesungen von E. Artin und E. Noether

5. Auflage der Modernen Algebra. 1967.
XII, 300 Seiten
(Heidelberger Taschenbücher, Band 23)
ISBN 3-540-03869-8

Inhaltsverzeichnis: Lineare Algebra. – Algebren. – Darstellungstheorie der Gruppen und Algebren. – Allgemeine Idealtheorie der kommutativen Ringe.– Theorie der Polynomideale. – Ganze algebraische Größen. – Bewertete Körper. – Algebraische Funktionen einer Variablen. – Topologische Algebra.

Aus den Besprechungen:
„Die Zahl der Studenten, welche aus den „Van der Waerden" Algebra gelernt haben, ist wohl kaum mehr abzuschätzen und wächst ins Unermäßliche, wenn man bedenkt, daß es unter den später erschienenen Algebra-Lehrbüchern kaum eines gibt, welches nicht durch diesen „Stammvater" aller modernen Algebra-Lehrbücher beeinflußt wurde. Wenn auch die Algebra in den etwa 30 Jahren seither sich beträchtlich ausgedehnt hat, ist das Werk auch heute noch keineswegs veraltet und braucht als Lehrbuch keine Konkurrenz zu scheuen…"

Int. Math. Nachrichten

Springer-Verlag
Berlin
Heidelberg
New York
Tokyo

Emmy Noether in Bryn Mawr

Proceedings of a Symposium
Sponsored by the Association of Women in
Mathematics in Honor of Emmy Noether's
100th Birthday

Editors: **B. Srinivasan, J. Sally**

1983. Approx. 16 figures.
Approx. 195 pages
ISBN 3-540-90838-2

Springer-Verlag
Berlin
Heidelberg
New York
Tokyo

Contents: Brauer Factor Sets, Noether Factor Sets and Crossed Products. – Noether's Problem in Galois Theory. – Noether Normalization. – Some Noncommutative Methods in Algebraic Number Theory. – Representations of Lie Groups and the Orbit Method. – Conservation Laws and Their Application in Global Differential Geometry.–Finite Simple Groups. – L^2-Cohomology and Intersection Cohomology of Certain Arithmetic Varieties. – Emmy Noether in Erlangen and Göttingen. – Emmy Noether in Bryn Mawr. – The Study of Linear Associative Algebras in the United States, 1870–1927. – Emmy Noether: Historical Contexts.